Andreas Graffunder
Rüdiger Hantsche
Irmfried Hartmann
Joerg Moebius
Zhiyun Ren
Matthias Boldin
Lutz Vietze

Contributions to
Autonomous Mobile Systems

Advances in Control Systems and Signal Processing
Editor: Irmfried Hartmann
Volume 7

Andreas Graffunder
Rüdiger Hantsche
Irmfried Hartmann
Joerg Moebius
Zhiyun Ren
Matthias Boldin
Lutz Vietze

Contributions to Autonomous Mobile Systems

With 65 Figures

vieweg

Editor:

Dr.-Ing. I. Hartmann
Prof. für Regelungstechnik und Systemdynamik
Technische Universität Berlin
Einsteinufer 17 – EN 11
1000 Berlin 10, West Germany

Printed on acid-free paper

ISBN 978-3-528-06383-2   ISBN 978-3-663-06842-6 (eBook)
DOI 10.1007/978-3-663-06842-6
ISSN   0724-9993

Contents

1 Introduction

Autonomous mobile systems (AMS) are systems capable of some mobility and equipped with advanced sensor devices in order to flexibly respond to changing environmental situations, thus achieving some degree of autonomy.

The purpose of this book is to contribute to some essential topics in this broad research area related to sensing and control, but not to present a complete design of an AMS.

Subjects concerning knowledge based control and decision, such as moving around obstacles, task planning and diagnosis are left for future publications in this series.

Research in the area of AMS has grown rapidly during the last decade, see e.g. [WAXMAN et al. 87], [DICKMANNS , ZAPP 87]. The requirements of an AMS strongly depends on the desired tasks the system should execute, its operational environment and the expected speed of the AMS.

For instance, road vehicles obtain velocities of 10 m/s and more, therefore the processing of sensor data such as video image sequences has to be very fast and simple, while indoor mobile robots deal with shorter distances and lower speeds, thus more sophistcated techniques are applicable and -as is done in our approach- additional sensors can be integrated to allow for multisensor processing.

The ideas presented in this book serve as guidelines to the realization of an autonomous mobile robot currently being built up at our institute. It mainly consists of a mobile platform with onboard mounted robotic manipulator, two stereo camera systems - one for navigating the platform, the other mounted on the endeffector of the manipulator-, and ultrasonic phased array sensors for modeling the closer environment. Parallel processing of the various sensor- and actuator data is executed by a multi-transputer system, offering the possibility of modular software design.

Chapter 2 gives an overview of robot modeling and -control and provides some basic issues needed in chapter 4.

Chapter 3 deals with processing of stereo image sequences for estimation of 3D-structure and relative movements between a moving observer and a moving object.

In chapter 4 the ideas introduced in chapter 3 are used in order to present a concept for visual robot control capable of handling moving objects.

Chapter 5 describes a possible approach to approximately observe the forces exerted on the endeffector of a robotic manipulator being in contact with its environment.

In chapter 6 are presented two concepts of airborne ultrasonic environment modeling using a phased array and multiple time of flight techniques.

Lane recognition and -following using monocular image sequences is the subject of chapter 7.

Finally, a concept of a multi-transputer system and its application to parallel processing is presented in chapter 8.

References

[DICKMANNS, ZAPP 87]
Dickmanns, ED.; Zapp, A.: "Autonomous High Speed Road Vechicle Guidance by Computer Vision"
Preprints of 10th IFAC-Congress, Vol. 4, München, 1987, pp. 232-237

[WAXMAN et al. 87]
Waxman, A.M.; Lemoigne, J.J.; Davis, L.S.; Srinivasan, B; Kushner, T.R.; Liang, E.: "A Visual Navigation System for Autonomous Land Vehicles"
IEEE Journal of Robotics and Automation, Vol. RA-3, No.2, 1987, pp. 124-141

2 Modeling and Control of Robotic Manipulators

M. Boldin, A. Graffunder, Z. Ren

The application of robotic manipulators in industrial manufacturing has grown rapidly during the last decades. In some fields such as spot welding and spray painting the use of robots is very common since reliable and rather simple modeling and control techniques are available. Utilisation of heavy rigid robots with high gear ratios allows to use simple linear models and linear control techniques. As robot application became more popular in industry it proved to be desirable to let the robot do more sophisticated tasks such as assembly, arc welding, grinding a.s.o. and to increase the robots working speed. The solution of these problems as well as the increasing appearance of direct drive robots required to consider the nonlinear coupled structure of robot mechanics.

Thinking of flexible joints or link flexibilities, which occur when light-weight robots are used, even more problems need to be solved by elaborated modeling and control techniques, but these will not be commented on here.

In the following some basic facts, definitions and notations will be introduced that are useful for the understanding of this part of the book. However the chapter should not be understood as a complete introduction to robotics, but rather as a base for the following chapters. For a more detailed introduction to the field of robotics c.f. [PAUL 81], [CRAIG 86], [VUKOBRATOVIC, POTKONJAK 82], [VUKOBRATOVIC, STOKIC 82], [FU et al. 87].

2.1 Modeling of robotic manipulators

In the following some basic considerations about the kinematic and dynamic structure of robots will be made and the most common methods to derive dynamic models for robots will be outlined. Throughout the chapter 'robot' shall stand for a rigid robot having no closed kinematic chains.

A robot can be seen as a multi-body-structure consisting of n rigid bodies called links connected by n joints, which are mostly either rotational or prismatic, usually giving the robot n degrees of freedom. The state of this system is definitly described by the joint variables q_i and $\dot{q}_i$, which are either angles and angular velocities in the case of rotational joints or positions and velocities in the case of prismatic joints. The process of robot modeling can generally be divided into two steps: the kinematic modeling and the derivation of dynamic equations.

Since the main purpose for robot application is to execute tasks in a certain environment, the tasks have to be brought into relation with this environment. In most cases this is done by defining a certain world or reference coordinate system fixed in the environment. Then the tasks will be described with respect to this system. As the position and orientation of the robots endeffector (respectivly hand or tool) described in the reference coordinate system depends on the joint variables, first of all a relation, namely a transformation, between the joint positions and the reference system has to be found. This relation is known as the kinematic model of a robot. For certain robot designs this kinematic model is rather simple (e.g. a three degree of freedom robot with three prismatic joints intersecting in one point under 90 degrees, would result in a kinematic model where each joint position corresponds to a cartesian coordinate of the reference system, if the latter is chosen suitably), but considering anthropomorphic robots the transformation can be quite complicated. There are numerous propositions described in the literature aiming to facilitate the derivation of this transformation sometimes even using computer algebra or expert systems. Common to all is the starting point, where link coordinate systems are defined in some way and transformations between adjoining link coordinate systems are built up. There are mainly three methods to build up these transformations using matrices: Euler angles, Yaw-Pitch-Roll angles and the Denavit-Hartenberg notation. The latter is the most common in robot kinematic modeling since it offers the possibility to include translatational displacements of adjoining link coordinate systems in the resulting 4x4 transformation matrix. The final transformation of the endeffector position defined by joint variables and the reference coordinate system results by multiplication of the link-to-link transformations and can be very complicated thinking of six or more degrees of freedom robots. Although the transformation from joint positions to world coordinates is definite, it is not one to one giving rise to difficulties concerning path planning due to ambiguities, singularities and redundancies.

For control purposes it is necassary to obtain knowledge of the dynamic behaviour of the robot, namely to derive a dynamic model. Considering a rotational joint i in an open kinematic chain as an example, it is obvious that the dynamics of this joint will be described by a second order differential equation. The inertia of the joint i depends on the configuration of the robot, namely the values of the following joint variables. Furthermore, accelerations of the other joints as well as centrifugal and coriolis effects due to motion of the other joints and gravitation will cause moments acting on joint i. Since the robot shall be controlled in some way, there will be actuators supplying forces or moments to change the joint positions. From these physical considerations the general dynamic model of a robot takes on the form:

$$\underline{M}(q)\,\ddot{q} + \underline{h}(q,\dot{q}) + \underline{g}(q) = \underline{\tau}. \qquad (2\text{-}1)$$

Here q denotes the n vector of joint variables, $\underline{M}(q)$ the symmetric positive definite nxn position dependent inertia matrix, $\underline{h}(q,\dot{q})$ the n vector describing centrifugal and coriolis forces, $\underline{g}(q)$ the n vector of gravitational forces and $\underline{\tau}$ the n vector of actuator forces.

If additionally Coulomb friction in the joints and external forces at the endeffector are introduced, the model can be expanded:

$$\underline{M}(q)\,\ddot{q} + \underline{h}(q,\dot{q}) + g(q) + \underline{B}\,\dot{q} + \underline{J}^T\,\underline{F}_e = \underline{\tau}\,, \tag{2-2}$$

where $\underline{B}$ denotes the nxn diagonal matrix of Coulomb friction, $\underline{J}$ the nxn jacobian matrix and $\underline{F}_e$ an n vector of external forces.

This model can be rewritten in state-space form:

$$\underline{x_1} = \underline{q}\,,\underline{x_2} = \underline{\dot{q}}$$

$$\underline{\dot{x}_1} = \underline{x_2}$$

$$\underline{\dot{x}_2} = \underline{M}^{-1}(\underline{x_1})\left[\,\underline{\tau} - \underline{h}(\underline{x_1},\underline{x_2}) - g(\underline{x_1}) - \underline{B}\,\underline{x_2} - \underline{J}^T\,\underline{F}_e\,\right]. \tag{2-3}$$

The problem that remains is the derivation of the matrices in (2-1) respectively (2-2). This can be done by the well known methods for mechanical multibody systems based e.g. on Newton/Euler or Lagrangian mechanics.

Newton/Euler methods consider forces and moments at the links and are based on the following formulas:

$$\underline{f}_i = m_i\,\ddot{r}_i$$

$$\underline{n}_i = \underline{I}_i\,\underline{\dot{\omega}}_i + \underline{\omega}_i \times \underline{I}_i\,\underline{\omega}_i \tag{2-4}$$

where $\underline{f}_i$ stands for the net force at link i, m_i for the mass of link i, $\ddot{r}_i$ for the acceleration of the center of gravity of link i, n_i for the net moment at link i, $\underline{I}_i$ for the inertia of link i and $\underline{\omega}_i$ for the angular velocity of link i.

Lagrangian methods consider the total energy of the robot and require the evaluation of the formula

$$\tau_i = \frac{d}{dt}\left(\frac{\partial L}{\partial \dot{q}_i}\right) - \frac{\partial L}{\partial q_i}\,;\, L = K - P;\, i = 1\, \dots\, n, \tag{2-5}$$

where the Lagrangian function L is the difference between kinetic energy K and potential energy P, q_i, $\dot{q}_i$ stands for the joint position and velocity respectively, and τ_i denotes the net forces/moments at link i.

Since it can be quite difficult to solve the problem of modeling by hand, recursive algorithms have been developed to simplify the derivation, e.g. [VUKOBRATOVIC, KIRCANSKI 85]. With the improvement of computer algebraic methods it has become possible to build up closed form dynamic robot models with the aid of the computer. There are of course other methods to derive the dynamic equations (e.g based on generalized d'Alembert equations of motion or Kane's equation) .

A three DOF example

As an example for the derivation of a dynamic model, a manipulator with three parallel rotational joints as illustrated in fig. 2-1 is considered. Here q_i, $\dot{q}_i$, $i=1\ldots3$ denote the joint angles and velocities, the link masses m_i, $i=1\ldots3$ are considered as point masses located in the middle of the links, l_i, $i=1\ldots3$ denote the link lenghts.

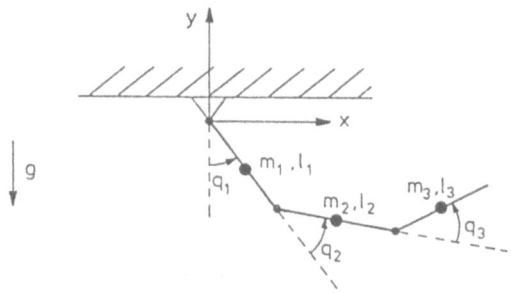

Fig. 2-1 Three link planar manipulator

We use the Lagrangian approach to derive explicit expressions for the elements of the matrices $\underline{M}(q)$, $\underline{h}(q , \dot{q})$, $\underline{g}(q)$ of the model (2-1), assuming that external forces do not exist and friction is negligible. In order to compute the Lagrangian function as stated in (2-5) expressions for the kinetic and potential energy have to be established. This is done by examining the individual links, where k_i and p_i, $i=1\ldots3$ are the kinetic and potential energies of the links. In the following J_{ip}, $i=1\ldots3$, $p=x, y, z$ denote the inertias for link i around axis p, v_i, $i=1\ldots3$ the cartesian velocities of the center of mass and $C_i = cos(q_i)$, $S_i = sin(q_i)$, $C_{ij} = cos(q_i + q_j)$, $S_{ij} = sin(q_i + q_j)$, $C_{ijk} = cos(q_i + q_j + q_k)$, $S_{ijk} = sin(q_i + q_j + q_k)$.

For link 1 it is seen that

$$k_1 = \frac{1}{2} m_1 v_1^2 + \frac{1}{2} J_{1z} \dot{q}_1^2,$$

$$= \frac{1}{2} m_1 \frac{l_1^2}{4} \dot{q}_1^2 + \frac{1}{2} J_{1z} \dot{q}_1^2,$$

$$p_1 = -m_1 g \frac{l_1}{2} C_1.$$

The evaluation of the kinetic and potential energy for links 2 and 3 is slightly more difficult, because the link velocities and angular velocities as well as the potential energy depend on the preceeding joints. Therefore only the results are given:

$$k_2 = \frac{1}{2} m_2 v_2^2 + \frac{1}{2} J_{2z} (\dot{q}_1 + \dot{q}_2)^2,$$

$$p_2 = -m_2 g (l_1 C_1 + \frac{l_2}{2} C_{12}),$$

$$k_3 = \frac{1}{2} m_3 v_3^2 + \frac{1}{2} J_{3z} (\dot{q}_1 + \dot{q}_2 + \dot{q}_3)^2,$$

$$p_3 = -m_3 g (l_1 C_1 + l_2 C_{12} + \frac{l_3}{2} C_{123}),$$

with

$$v_2^2 = l_1^2 \dot{q}_1^2 + \frac{l_2^2}{4} (\dot{q}_1 + \dot{q}_2)^2 + l_1 l_2 \dot{q}_1 (\dot{q}_1 + \dot{q}_2) C_2,$$

$$v_3^2 = l_1^2 \dot{q}_1^2 + l_2^2 (\dot{q}_1 + \dot{q}_2)^2 + 2 l_1 l_2 \dot{q}_1 (\dot{q}_1 + \dot{q}_2) C_2 + \frac{l_3^2}{4} (\dot{q}_1 + \dot{q}_2 + \dot{q}_3)^2 +$$

$$+ l_3 (\dot{q}_1 + \dot{q}_2 + \dot{q}_3) \Big[l_1 \dot{q}_1 C_{23} + l_2 (\dot{q}_1 + \dot{q}_2) C_3 \Big].$$

Now the Lagrangian

$$L = K - P = \sum_{i=1}^{3} (k_i - p_i)$$

can be built up resulting in

$$L = \frac{1}{2} \{ (\frac{m_1}{4} + m_2 + m_3) \, l_1^2 \, \dot{q}_1^2 + (\frac{m_2}{4} + m_3) \, l_2^2 \, (\dot{q}_1 + \dot{q}_2)^2 +$$

$$+ (m_2 + 2 \, m_3) \, l_1 \, l_2 \, \dot{q}_1 \, (\dot{q}_1 + \dot{q}_2) \, C_2 + \frac{m_3}{4} \, l_3^2 \, (\dot{q}_1 + \dot{q}_2 + \dot{q}_3)^2 +$$

$$+ m_3 \, l_3 \, (\dot{q}_1 + \dot{q}_2 + \dot{q}_3) \, [\, l_1 \, \dot{q}_1 \, C_{23} + l_2 \, (\dot{q}_1 + \dot{q}_2) C_3 \,] +$$

$$+ J_{1z} \, \dot{q}_1^2 + J_{2z} \, (\dot{q}_1 + \dot{q}_2)^2 + J_{3z} \, (\dot{q}_1 + \dot{q}_2 + \dot{q}_3)^2 \} +$$

$$+ g \{ (\frac{m_1}{2} + m_2 + m_3) \, l_1 \, C_1 + (\frac{m_2}{2} + m_3) \, l_2 \, C_{12} + \frac{m_3}{2} \, l_3 \, C_{123} \}. \tag{2-6}$$

Considering the robot model (2-1) and eq. (2-5) the elements of the matrices $\underline{M}(q)$, $\underline{C}(q, \dot{q})$, $g(q)$ will be calculated. Partial derivation of (2-6) with respect to $\dot{q}_1$ and derivation with respect to time, respectively partial derivation with respect to q_1 yields

$$\frac{d}{dt} \left\{ \frac{\partial L}{\partial \dot{q}_1} \right\} = \frac{1}{2} \{ 2 \, (\frac{m_1}{4} + m_2 + m_3) \, l_1^2 \, \ddot{q}_1 + 2 \, (\frac{m_2}{4} + m_3) \, l_2^2 \, (\ddot{q}_1 + \ddot{q}_2) -$$

$$- (m_2 + 2 \, m_3) \, l_1 \, l_2 \, S_2 \, \dot{q}_2 \, (2\dot{q}_1 + \dot{q}_2) + (m_2 + 2m_3) \, l_1 \, l_2 \, C_2 \, (2\ddot{q}_1 + \ddot{q}_2) +$$

$$+ \frac{m_3}{2} \, l_3^2 \, (\ddot{q}_1 + \ddot{q}_2 + \ddot{q}_3) +$$

$$+ m_3 \, l_3 \, [l_1 \, \ddot{q}_1 \, C_{23} - l_1 \, \dot{q}_1 \, S_{23} \, (\dot{q}_2 + \dot{q}_3) + l_2 \, (\ddot{q}_1 + \ddot{q}_2) \, C_3 - l_2 \, (\dot{q}_1 + \dot{q}_2) \, S_3 \, \dot{q}_3] +$$

$$+ 2 \, J_{1z} \, \ddot{q}_1 + 2 \, J_{2z} \, (\ddot{q}_1 + \ddot{q}_2) + 2 \, J_{3z} (\ddot{q}_1 + \ddot{q}_2 + \ddot{q}_3) +$$

$$+ m_3 \, l_3 \, (\ddot{q}_1 + \ddot{q}_2 + \ddot{q}_3) \, (l_1 \, C_{23} + l_2 \, C_3) -$$

$$- m_3 \, l_3 \, (\dot{q}_1 + \dot{q}_2 + \dot{q}_3) \, (l_1 \, S_{23} \, (\dot{q}_2 + \dot{q}_3) + l_2 \, S_3 \, \dot{q}_3) \},$$

$$\frac{\partial L}{\partial q_1} = -g \, [\, (\frac{m_1}{2} + m_2 + m_3) \, l_1 \, S_1 + (\frac{m_2}{2} + m_3) \, l_2 \, S_{12} + \frac{m_3}{2} \, l_3 \, S_{123} \,].$$

Sorting terms depending on $\ddot{q}_1$, $\ddot{q}_2$, $\ddot{q}_3$, the gravitational constant g and all other terms results in the matrix elements:

$$M_{11}(q) = (\frac{m_1}{4} + m_2 + m_3)\, l_1^2 + (\frac{m_2}{4} + m_3)\, l_2^2 + \frac{m_3}{4}\, l_3^2 +$$

$$+ (m_2 + 2\,m_3)\, l_1\, l_2\, C_2 + m_3\, l_3\, (l_1\, C_{23} + l_2\, C_3) +$$

$$+ J_{1z} + J_{2z} + J_{3z},$$

$$M_{12}(q) = (\frac{m_2}{4} + m_3)\, l_2^2 + \frac{m_3}{4}\, l_3^2 +$$

$$+ (\frac{m_2}{2} + m_3)\, l_1\, l_2\, C_2 + \frac{m_3}{2}\, l_3\, (l_1\, C_{23} + 2\, l_2\, C_3) +$$

$$+ J_{2z} + J_{3z},$$

$$M_{13}(q) = \frac{m_3}{4}\, l_3^2 + \frac{m_3}{2}\, l_3\, (l_1\, C_{23} + l_2\, C_3) + J_{3z},$$

$$h_1(q, \dot{q}) = -(\frac{m_2}{2} + m_3)\, l_1\, l_2\, S_2\, \dot{q}_2(2\,\dot{q}_1 + \dot{q}_2) -$$

$$- \frac{m_3}{2}\, l_3\left[l_1\, \dot{q}_1 S_{23}\, (\dot{q}_2 + \dot{q}_3) + l_2\, (\dot{q}_1 + \dot{q}_2)\, S_3\, \dot{q}_3 \right] -$$

$$- \frac{m_3}{2}\, l_3\, (\dot{q}_1 + \dot{q}_2 + \dot{q}_3)\left[l_1\, S_{23}\, (\dot{q}_2 + \dot{q}_3) + l_2\, S_3\, \dot{q}_3 \right],$$

$$g_1(q) = g\,[(\frac{m_1}{2} + m_2 + m_3)\, l_1\, S_1 + (\frac{m_2}{2} + m_3)\, l_2\, S_{12} + \frac{m_3}{2}\, l_3\, S_{123}].$$

Similiarly one obtains the other elements by setting i=2 and i=3 in (2-5). The differentiation is omitted for brevity and only the results are given (remind that $\underline{M}(q)$ is symmetric):

$$M_{21}(q) = M_{12}(q),$$

$$M_{22}(q) = (\frac{m_2}{4} + m_3)\, l_2^2 + \frac{m_3}{4}\, l_3^2 + m_3\, l_2\, l_3\, C_3 + J_{2z} + J_{3z},$$

$$M_{23}(q) = \frac{m_3}{4} l_3^2 + \frac{m_3}{2} l_2 l_3 C_3 + J_{3z},$$

$$M_{31}(q) = M_{13}(q),$$

$$M_{32}(q) = M_{23}(q),$$

$$M_{33}(q) = \frac{m_3}{2} l_3^2 + J_{3z},$$

$$h_2(q, \dot{q}) = -\frac{m_3}{2} l_3 \left[-l_1 \dot{q}_1^2 S_{23} + l_2 (\dot{q}_1 + \dot{q}_2) \dot{q}_3 S_3 + (\dot{q}_1 + \dot{q}_2 + \dot{q}_3) \dot{q}_3 l_2 S_3 \right] +$$

$$+ (\frac{m_2}{2} + m_3) l_1 l_2 \dot{q}_1^2 S_2,$$

$$h_3(q, \dot{q}) = \frac{m_3}{2} l_3 \left[l_1 \dot{q}_1^2 S_{23} + l_2 (\dot{q}_1 + \dot{q}_2)^2 S_3 \right],$$

$$g_2(q) = g \left[(\frac{m_2}{2} + m_3) l_2 S_{12} + \frac{m_3}{2} l_3 S_{123} \right],$$

$$g_3(q) = g \frac{m_3}{2} l_3 S_{123}.$$

This completes the derivation of the 3 DOF planar robot model and shows the complexity of robot modeling by hand.

2.2 Control of robotic manipulators in joint space

Based on the standard dynamic model for a robot with n degrees of freedom (2-1) - (2-3) the most popular control techniques in the joint level will be outlined here. The purpose of robot control depends on the task that shall be performed.

- In point-to-point movements control shall assure a certain position accuracy.

- In path tracking tasks control shall eliminate the error between the actual and the reference trajectory.

- In some fields of assembly besides position control, force control is needed to assure a certain contact force between the robots endeffector and the environment during interaction, e.g. screwing, inserting. This will be discussed in section 2.4.

Due to these different purposes, different control techniques have been developed, that should be outlined in the following.

As mentioned before the use of heavy rigid robots with high gear ratios is common in industrial practice, because of their reliability and robustness. Furthermore the joint velocities in most of the industrial applications are rather small (less than 1 m/s). Therefore coupling as well as centrifugal and coriolis effects can usually be neglected and the problem of controlling the nonlinear multivariable system 'robot' is simplified, since only the actuator dynamics remain to be considered. Thus the robot can be seen as a system of n decoupled linear timeinvariant differential equations allowing to use all the well known techniques from linear control theory. Consideration of changing payload or slight coupling leads to some kind of feedforward control, i.e. compensation, but conserves the linear structure. The most popular structure is a cascaded current, velocity, position control with feedforward compensation of gravitational forces if necessary.

If higher joint velocities or smaller gear ratios are used neglection of coupling and nonlinear effects may lead to poor performance, limit cycles or even instability. A linearization along the desired trajectory may help in solving these problems and methods for linear timevariable systems may be applied.

However if changing tasks, i.e. changing trajectories, have to be performed even this approach fails and nonlinear methods have to be introduced. The aim of most of the nonlinear techniques can be stated as follows:

Find a compensating control law, possibly using measurements of joint magnitudes, i.e. position, velocity, acceleration, that leaves the nonlinear multivariable system 'robot' as a linear decoupled system.

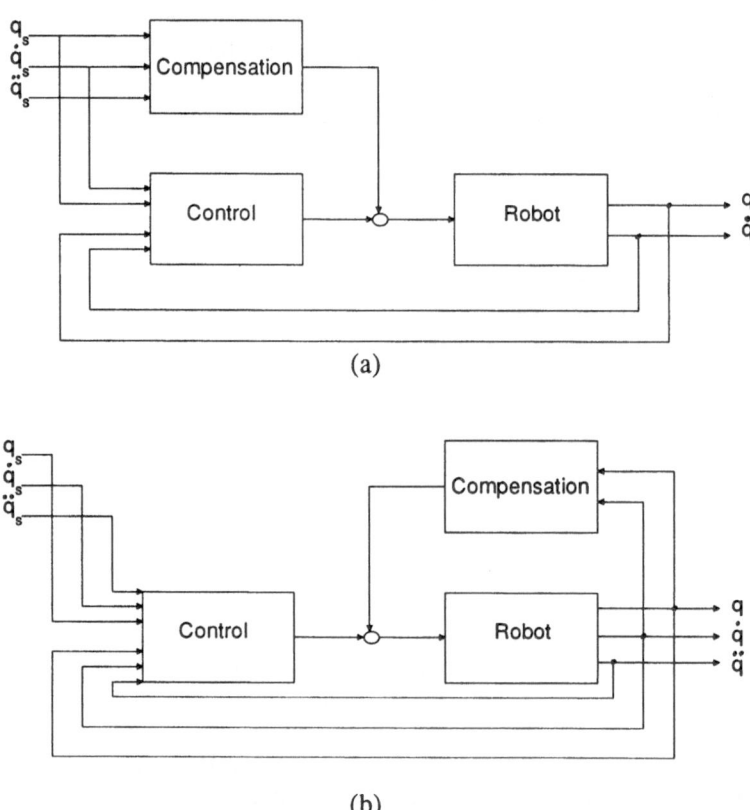

Fig. 2-2 Decoupling control: (a) Feedforward compensation
(b) Feedback compensation

Basically there are two possible structures: feedforward and feedback compensation (fig. 2-2). In feedforward compensation the known reference values of joint position, velocity and acceleration are used to derive an actuating force/torque that compensates all nonlinear effects using a nominal robot model. An outer control loop assures errors in position or velocity to be driven to zero. In feedback compensation measurements of joint position and velocity are used for the same purpose. Well known methods of this kind are e.g. 'nonlinear decoupling' [FREUND 75], [FREUND 82], 'dynamic control' [BEJCZY, TARN 88], 'computed torque' (e.g. [FU et al. 87]).

Another method for deriving the same compensation law can be introduced applying

Ljapunov's second method, see e.g. [BÖCKER, HARTMANN, ZWANZIG 86]. Given the nonlinear state space model for the joint dynamics (2-3), considering only position control and thus neglecting external forces, the following positive definite function maybe introduced as a Ljapunov function candidate:

$$V(\underline{x}_1, \underline{x}_2) = \frac{1}{2} \left(\underline{x}_1^T \ \underline{x}_2^T \right) \begin{pmatrix} \underline{\Lambda}_1 & \underline{0} \\ \underline{0} & \underline{E}_n \end{pmatrix} \begin{pmatrix} \underline{x}_1 \\ \underline{x}_2 \end{pmatrix} \tag{2-7}$$

where $\underline{\Lambda}_1 = diag\,(\,\lambda_{1i}\,)$, $i = 1...n$, $\lambda_{1i} > 0$, and $\underline{E}_n$ is the nxn identity matrix. Derivation of (2-7) using (2-3) leads to

$$\dot{V}(\underline{x}_1, \underline{x}_2) = \underline{x}_2^T \,[\, \underline{\Lambda}_1 \underline{x}_1 + \dot{\underline{x}}_2]$$

$$= \underline{x}_2^T \left\{ \underline{\Lambda}_1 \underline{x}_1 + \underline{M}^{-1}(\underline{x}_1) \,[\, \underline{\tau} - \underline{h}(\underline{x}_1, \underline{x}_2) - \underline{g}(\underline{x}_1) - \underline{B}\,\underline{x}_2 \,] \right\}, \tag{2-8}$$

which can be made negative semidefinite using the decoupling control law

$$\underline{\tau}_d = \underline{h}(\underline{x}_1, \underline{x}_2) + \underline{g}(\underline{x}_1) + \underline{B}\,\underline{x}_2 - \underline{M}(\underline{x}_1) \,[\, \underline{\Lambda}_1 \underline{x}_1 + \underline{\Lambda}_2 \underline{x}_2 \,],$$

$$\underline{\Lambda}_2 = diag\,(\,\lambda_{2i}\,),\ i = 1\ ...\ n,\ \lambda_{2i} > 0. \tag{2-9}$$

It should be obvious, that the negative semidefiniteness is sufficient for asymptotic stability, since $\underline{x}_2 = \underline{0}$, $\underline{x}_1 = const. \neq \underline{0}$ is no solution of the autonomous differential equation. Substitution of (2-9) into (2-3) results in

$$\dot{\underline{x}}_1 = \underline{x}_2$$

$$\dot{\underline{x}}_2 = -\underline{\Lambda}_1 \underline{x}_1 - \underline{\Lambda}_2 \underline{x}_2.$$

If the states are rearranged in the kind that corresponding joint positions and velocities are brought together it is seen, that the state space model of the decoupled system takes on the form

$$
\begin{pmatrix} \dot{x}_{11} \\ \dot{x}_{21} \\ \dot{x}_{12} \\ \dot{x}_{22} \\ \cdots \\ \dot{x}_{1n} \\ \dot{x}_{2n} \end{pmatrix} = \begin{pmatrix} 0 & 1 & 0 & 0 & \cdots & 0 & 0 \\ -\lambda_{11} & -\lambda_{21} & 0 & 0 & \cdots & 0 & 0 \\ 0 & 0 & 0 & 1 & \cdots & 0 & 0 \\ 0 & 0 & -\lambda_{12} & -\lambda_{22} & \cdots & 0 & 0 \\ \cdots & \cdots & \cdots & \cdots & \cdots & \cdots & \cdots \\ 0 & 0 & 0 & 0 & \cdots & 0 & 1 \\ 0 & 0 & 0 & 0 & \cdots & -\lambda_{1n} & -\lambda_{2n} \end{pmatrix} \begin{pmatrix} x_{11} \\ x_{21} \\ x_{12} \\ x_{22} \\ \cdots \\ x_{1n} \\ x_{2n} \end{pmatrix} . \qquad (2\text{-}10)
$$

Defining

$$
\underline{z}_i := \begin{pmatrix} x_{1i} \\ x_{2i} \end{pmatrix}
$$

and

$$
\underline{A}_i = \begin{pmatrix} 0 & 1 \\ -\lambda_{1i} & -\lambda_{2i} \end{pmatrix}
$$

(2-10) can be written

$$
\begin{pmatrix} \dot{\underline{z}}_1 \\ \dot{\underline{z}}_2 \\ \cdots \\ \dot{\underline{z}}_n \end{pmatrix} = \begin{pmatrix} \underline{A}_1 & 0 & \cdots & 0 \\ 0 & \underline{A}_2 & \cdots & 0 \\ \cdots & \cdots & \cdots & \cdots \\ 0 & 0 & \cdots & \underline{A}_n \end{pmatrix} \begin{pmatrix} \underline{z}_1 \\ \underline{z}_2 \\ \cdots \\ \underline{z}_n \end{pmatrix} . \qquad (2\text{-}11)
$$

In order to control the individual joints additional input forces/moments $\underline{\tau}_c := \underline{M}\,\underline{w}$ maybe added to the decoupling forces/moments $\underline{\tau}_d$. Thus we finally have the linear decoupled manipulator:

$$
\dot{\underline{z}} = \underline{A}\,\underline{z} + \underline{w}. \qquad (2\text{-}12)
$$

This approach is applied in chap.4.

The techniques mentioned so far assume perfect knowledge about the manipulator dynamics, that means they do not take into account model inaccuracies due to e.g. changing payload or friction etc. In order to cope with these problems adptive control has been proposed, see e.g. [CRAIG 88]. Starting from a linearization around a preplanned nominal trajectory, known techniques like MRAC or Self-Tuning-Control are applied.

Numerous other methods of joint space control may be found in the literature, including predictive and sliding mode control, e.g. [YOUNG 78], [KUNTZE et al. 88], that shall not be commented on, since this short introduction does not claim to be exhaustive.

2.3 Control of robotic manipulators in cartesian space

In many applications it is natural to define the task of manipulator control in terms of a trajectory that specifies a desired motion of the endeffector. In order to specify the motion trajectory completely, the time history of desired position and orientation of a coordinate system attached to the endeffector with respect to the cartesian world coordinate system (called "pose") have to be defined. Representing the position by a vector that points from the origin of the world coordinate system to the endeffector coordinate system and the orientation by a rotation matrix, it is conceptually simple to calculate the corresponding joint positions and do the manipulator control in joint space. But as mentioned earlier, the required inverse transformation may pose severe difficulties since generally it does not have unique solutions, even if the manipulator has no redundant degrees of freedom. Alternatively, the desired motion trajectory may be equivalently specified by defining the time derivative of the desired position $\underline{r}_o$ and the desired angular velocity $\underline{\omega}_c$ of the endeffector coordinate system over time. These are linked to the joint velocities by a linear mapping involving the Jacobian matrix of the manipulator:

$$\underline{\dot{s}}(t) = \begin{pmatrix} \underline{\dot{r}}_o{}'(t) \\ \underline{\omega}_c(t) \end{pmatrix} = \underline{J}(\underline{q}) \, \underline{\dot{q}}(t). \tag{2-13}$$

The desired trajectory of the joint velocities $\dot{q}_d(t)$ may be uniquely determined from the cartesian coordinates by applying the inverse of $\underline{J}(\underline{q})$, provided $\underline{J}(\underline{q})$ is regular. In case that the manipulator has redundant degrees of freedom the generalized inverse may be used. Unfortunately even if no redundancies are present, there are singular configurations in joint space where one or more degrees of freedom are lost rendering $\underline{J}(\underline{q})$ not invertible. This is an inherent problem of manipulator control and must be taken into account in the task-specification phase. Applying one of the methods mentioned in the last section, the joint velocities may be controlled according to the calculated reference trajectory $\dot{q}_d(t)$. This approach has been introduced by [WHITNEY 69] and is called "resolved motion rate control " (RMRC). It has been extended to include specifications of desired accelerations (called "resolved motion acceleration control", RMAC, by [LUH et al. 80]). These techniques overcome the difficulties posed by the inverse kinematic transformation of reference values defined in cartesian space into those defined in joint space. However two deficiencies of these concepts remain. First, since the actual control is done in joint space, errors in the actual pose of the endeffector due to elasticity, backlash in the gears and little inaccuracies in the geomatrical model of the manipulator are not measured and therefore can not be suppressed.

Secondly, it always has been assumed that a preplanned trajectory in cartesian space is available. Of course this requirement severely limits applications of these techniques with regard to enhanced flexibility of robotic manipulators. In order to achieve some degree of autonomy the manipulator control should be adaptable to changing situations of its environment.

There is some hope that these difficulties may at least be partially overcome by incorporating powerful sensing devices operating in cartesian space into the manipulator control loop. This is the topic of chapters 3 and 4, where concepts relying on 3D-vision are presented.

2.4 Force control of robotic manipulators

To enable the robot to complete tasks in contact with the environment assuring desired contact forces, requires the measurement and control of these contact forces. The theory of the so called 'force control' or 'compliant control' is based mainly on the research in the fields of telemanipulation and artificial arms. The problem that arises is that slight position errors due to programming, task planning or insufficient control, that always occur, can make it impossible to e.g. insert a peg in a hole due to misalignments and angular errors.

Besides 'direct force control' that uses a desired force instead of a position vector other techniques are available. Some of them vary the robots stiffness by changing damping parameters of the control laws ('stiffness control', 'damping control', 'impedance control') and adding a certain force reference variable. A more task oriented method is 'hybrid control', where certain degrees of freedom in global coordinates can be switched to be either position or force controlled depending on the respective task (e.g. in contour following the degrees of freedom belonging to the normal vector of the surface are force controlled while the others are position controlled). A good overview of force control techniques was given by [WHITNEY 87].

For all these methods force measurements are needed, that can be done by force/torque sensors located in the robots hand. Since these sensors are still rather expensive considering small robots, it is tried to use estimation and observation techniques. A possible approach to approximately observe endeffector forces is given in chapter 5.

There are also passive approaches like the 'remote center compliance' using a flexible connection between robot arm and endeffector, enabling the endeffector to correct slight position errors without active actuator commands, or special programming or task planning techniques, e.g. 'wobbling' a peg in a hole.

Finally it has to be mentioned that most of the sophisticated control algorithms are only used in laboratory robot control and only a few industrial applications exist because of the enormous computing requirements. If computers become even more powerful in the following years e.g. by parellel architectures it can be thought of acceptance by industrial users.

2.5 References

[BEJCZY, TARN 88]
Bejczy, A. K. ; Tarn, T. J. : "Dynamic Control of Robot Arms in Task Space Using Nonlinear Feedback"
Automatisierungstechnik, vol. 36, no. 10, 1988

[BÖCKER, HARTMANN, ZWANZIG 86]
Böcker, J. ; Hartmann, I. ; Zwanzig, C. : "Nichtlineare und udaptive Regelungssysteme"
Springer Verlag, Berlin, 1986

[CRAIG 86]
Craig, J. J. : "Introduction to Robotics: Mechanics and Control"
Addison-Wesley, Reading, Mass. , 1986

[CRAIG 88]
Craig, J. J. : "Adaptive Control of Mechanical Manipulators"
Addison-Wesley, Reading, Mass. , 1988

[FREUND 75]
Freund, E. : "The structure of decoupled non-linear systems"
Int. J. Control, vol. 21, no. 3, pp. 443-450, 1975

[FREUND 82]
Freund, E. : "Fast Nonlinear Control with Arbitrary Pole Placement for Industrial Robots and Manipulators"
Int. J. Robotics Res. , vol. 1, no. 1, pp. 65-78, 1982

[FU et al. 87]
Fu, K. S. ; et al. : "Robotics: Control, Sensing, Vision and Intelligence"
McGraw-Hill Book Company, New York, 1987

[KUNTZE et al 88]
Kuntze, H.-B. : "On the Application of a New Method for Fast and Robust Position Control of Industrial Robots"
Proc. IEEE Int. Conf. on Robotics and Automation, Philadelphia, 1988

[LUH et al. 80]
Luh, J. Y. S. ; et al. : "Resolved-acceleration Control of Mechanical Manipulators"
Trans. ASME, J. Dynamic Systems, Measurement and Control, vol. 120, pp. 69-76, 1980

[PAUL 81]
Paul, R. P. : "Robot Manipulator: Mathematics, Programming and Control"
MIT Press, Cambridge, Mass., 1981

[VUKOBRATOVIC, POTKONJAK 82]
Vukobrativic, M. ; Potkonjak, V. : "Dynamics of Manipulation Robots"
Springer Verlag, Berlin, 1982

[VUKOBRATOVIC, STOKIC 82]
Vukobratovic, M. ; Stokic, D. : "Control of Manipulation Robots"
Springer Verlag, Berlin, 1982

[VUKOBRATOVIC, KIRCANSKI 85]
Vukobratovic, M. ; Kircanski, N. : "Real-Time Dynamics of Manipulation Robots"
Springer Verlag, Berlin, 1985

[WHITNEY 69]
Whitney, D. E. : "Resolved Motion Rate Control of Manipulators and Human Prostheses"
IEEE Trans. Man-Machine Systems, vol. MMS-10, no. 2, pp. 47-53, 1969

[WHITNEY 87]
Whitney, D. E. : "Historical Perspectives and State of the Art in Robot Force Control"
Int. J. Robotics Res. , vol. 6, no. 1, pp. 3-14, 1987

[YOUNG 78]
Young, K. K. D. : "Controller Design for a Manipulator Using Theory of Variable Structure Systems"
IEEE Trans. Systems, Man, Cybern. , vol. SMC-8, no. 2, pp. 101-109, 1978

3 Estimation of Structure and Relative Motion from Stereo Image Sequences

A. Graffunder

3.1 Introduction

One of the key-problems in realizing Autonomous Mobile Systems (AMS) is deriving descriptions of the structure and motions of objects in the environment of an AMS.

A mobile platform for example, needs to know the geometry, position and orientation as well as the velocities of a possibly moving obstacle in order to avoid it.

The same problem relates to vision based control schemes for robotic manipulators, where the manipulator joints are controlled for reaching, grasping and manipulating in closed visual loop. The geometry, position and orientation of the workpiece and their changes need to be determined by exploiting information from visual sensors that are mounted either stationary, w.r.t. the base of the robot, or directly on the endeffector ("eyes in hand technique"), see chapter 4.

The common feature of these examples is that movements of an agent (platform, endeffector etc.) with attached coordinate frame are realized relative to other possibly moving objects of the environment.

Inferring the 3D-structure and relative movements of the environment from 2D-images has been studied since the late seventies. Usually, the structure was defined by the 3D coordinates of some points attached to a rigid object, sometimes called "prominent points".

Since the third dimension is lost by projection of 3D-points into 2D-image planes, the 3D-structure can not be inferred from a single view. This inherent problem can be overcome by taking images of the same object from different view-points, thus creating an image sequence. Several ways to put this idea into reality have been suggested, differing in the assumptions imposed on image taking, possible observer movements and movements of scene-objects.

Conceptually, the simplest approach to 3D-structure determination is stereoscopy, taking two views from different, known camera positions and determining the 3D-position of structure points by triangulation, see e.g. [DUDA, HART 73].

Another approach is to consider a single camera, moving with respect to a stationary rigid object (which can be the whole environment), taking pictures from different unknown positions. A number of image features, related to points of the environment are tracked in successive images of the resulting monocular image sequence, where often only very short sequences are used (two or three images). Although 3D-structure determination often has been of primary interest, the camera-movements need to be identified too, since the two problems are interrelated. Applying the laws of perspective transformations, a set of algebraic nonlinear equations can be derived, relating the image coordinates of tracked features to the 3D-coordinates of the observed structure points and camera-displacements in terms of translations and rotations. Provided that enough points are observed, the structure coordinates and motion parameters can be determined. However, as a consequence of using only one camera, the structure and translations are only unique up to a common scale factor.

[ROACH, AGGARWAL 80] used five points in two views resulting in a set of 20 nonlinear equations in 20 unknowns which were solved using an iterative Levenberg-Marquardt algorithm.

[NAGEL, NEUMANN 81] separated the problem by first solving a set of fourth order polynomials for the rotation parameters and then calculating structure and translation parameters.

[TSAI, HUANG 84] presented a linear approach, however, for achieving linearity, a minimum number of eight points have to be observed.

Since the localization of image features is always corrupted by noise, results may become poor if the minimum number of points is used. The noise sensitivity can be reduced by spatial aggregation, i.e. by observing more points and solving the resulting overdetermined equations by using least squares fitting techniques. This may cause serious problems when the information contained in the images is not rich enough for tracking many structure points. Furthermore, since in many applications longer image sequences are available, it is not clear how estimations from successive short sub-sequences should be combined in order to get improved estimates over time.

A second possibility of noise reduction, namely by temporal aggregation is the idea underlying model-based approaches relying on stochastic filtering theory. Relative movements are modeled by the solutions of differential equations, reflecting the fact, that the velocities of an object with finite mass change smoothly over time.

[WU et al. 88] presented a technique based on an extended Kalman filter (EKF) suitable for movement estimation. Assuming that the camera as well as the observed object may be moving , they were able to estimate the true 3D-position and orientation and the translational and rotational velocities of the object relative to the camera. However, in order to overcome the ambiguities caused by using only a single camera, they required the structure of the observed object to be known. The 3D-coordinates of at least six object-points have to be known w.r.t. an object centered coordinate frame.

Since this assumption is often too restrictive, the evaluation of stereo-image sequences has been investigated, offering the possibility to decouple the problem of movement estimation from structure estimation.

[HUANG, BLOSTEIN 85] considered 3D-motion estimation from two successive stereo-pairs. Knowing the 3D-coordinates of only three non-coplanar object points from a stereo triangulation process, the 3D-motion of the object can be determined uniquely. However, with only three points the authors reported poor results, caused by the fact that the errors in calculating the depth of a 3D-point by stereo triangulation increase quadratically with the distance of the point from the cameras of the stereo setup. Once again the problem could be alleviated by spatial aggregation. Using measurements from nine observed points in connection with a heuristically motivated, iterative weighted least squares technique, the structure and motion parameters could be estimated fairly accurately.

[YOUNG, CHELLAPPA 88] considered the estimation of position and orientation as well as translational and rotational velocities of an object moving w.r.t. a stationary stereo setup using an EKF. Here again, tracking three object-points is sufficient to determine the desired quantities. As a result of temporal aggregation, the estimates converge to the true values with increasing length of the stereo-image sequence.

Since for applications in the context of AMS a high degree of flexibility is desired, assumptions about a priori knowledge concerning the 3D-structure and relative motions must be kept at a minimum. Therefore, in this work stereo vision combined with an EKF approach is investigated. The detailed problem formulation suitable for AMS is given in the next section.

3.2 Problem statement

In this work, we treat the case where the visual sensing system is mounted directly on the AMS, so that the coordinate frame attached to the AMS is identical with the coordinate frame of the visual sensing system, called "observer coordinate frame" in the sequel. Other cases, for example a robot manipulator with stationary mounted vision system, may be treated in a similar way as is presented in this chapter.

The general situation to be examined here is depicted in fig. 3-1. There are three coordinate frames involved:

- the world coordinate frame (O,x,y,z), attached to the stationary environment of the AMS,

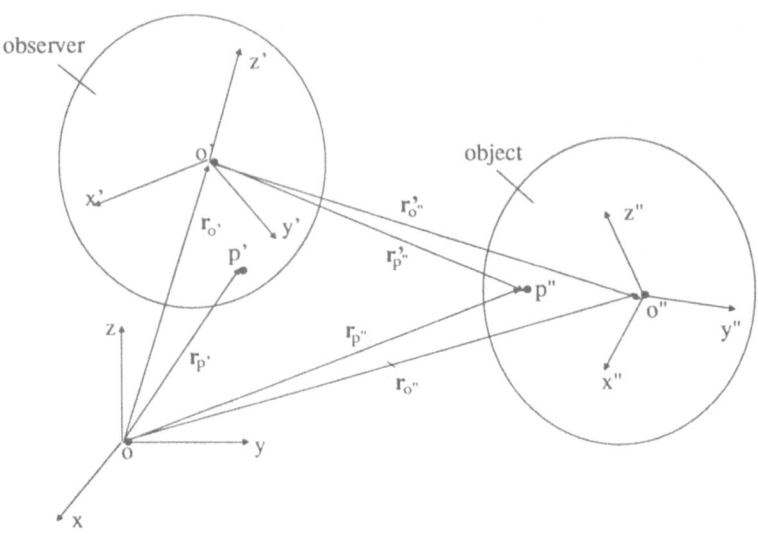

Fig.3-1 Coordinate frames of stationary environment,
moving observer and moving scene-object

- the observer coordinate frame (O',x',y',z'), attached to the moving observer, and

- the object coordinate frame (O",x",y",z"), attached to a scene-object of interest.

 Throughout this chapter, vector quantities with components defined w.r.t. the world-observer- or object-coordinate frame are denoted with no prime, one prime or two primes respectively.

Assuming P" is a fixed point of a moving object, it is well known that the velocity of P" is composed of a translational and a rotational component

$$\dot{\underline{r}}_{p"} = \dot{\underline{r}}_{o"} + \underline{\omega}_b \times (\underline{r}_{p"} - \underline{r}_{o"}) \tag{3-1}$$

where $\underline{r}_{o"}$ is a reference point on the object, assumed to be the origin of the object coordinate frame, and $\underline{\omega}_b$ denotes its angular velocity. $\dot{\underline{r}}_{o"}$ is commonly called "translational velocity".

A similar relation exists for a point P' rigidly attached to the observer:

$$\dot{\underline{r}}_{p'} = \dot{\underline{r}}_{o'} + \underline{\omega}_c \times (\underline{r}_{p'} - \underline{r}_{o'}) \tag{3-2}$$

where $\underline{r}_{o'}$ is the observer-origin and $\dot{\underline{r}}_{o'}$ and $\underline{\omega}_c$ denote the translational and angular velocty of the observer respectively.

Since we are interested in the relative movements between observer and object we require a similar formulation as (3-1), but with respect to coordinates of the observer. Formally we may write

$$\dot{\underline{r}}'_{p''} = \dot{\underline{r}}'_{o''} + \widetilde{\underline{\omega}} \times (\underline{r}'_{p''} - \underline{r}'_{o''}) \tag{3-3}$$

$\dot{\underline{r}}'_{p''}$ denoting the relative translational velocity and $\widetilde{\underline{\omega}}$ the relative angular velocity of the object w.r.t. the observer.

Now, the problem to be considered in the sequel can be stated as follows: How can

- the structure of an object of interest, represented by a set of prominent object points,

- its relative pose (i.e. its relative position and orientation)

- and its relative movements (in terms of $\dot{\underline{r}}'_{o''}$ and $\widetilde{\underline{\omega}}$)

be derived from visual data? In order to formulate a proper model based estimation problem, differential equations describing the evolutions of the desired quantities have to be derived. This is done in sec. 3.3 and 3.4.

An important requirement concerning the movement model to be derived is, that the dependencies of the relative velocities of the observer and object, i.e. $\dot{\underline{r}}'_{o''}$ and $\widetilde{\underline{\omega}}$ on their absolute velocities, i.e $\dot{\underline{r}}_{o'}$, $\dot{\underline{r}}_{o''}$ $\underline{\omega}_c$ and $\underline{\omega}_b$ are explicitely contained in the model. This is clear in view of the intended application, namely visual movement control, since for generating appropriate control signals for the observer kinematics, the impact of the absolute movements of the observer on relative movements must be known. A movement model that explains how relative movements are build up from the individual movements of observer and object is also of interest from a theoretical point of view, since it reveals what quantities are recoverable from certain visual measurements. This point is shortly dealt with in sec. 3.6. Visual measurements are assumed to be extracted from stereo-image sequences providing true 3D-information of the environment and therefore avoiding overly restricted assumptions on a priori knowledge of the scene, usually required by techniques using monocular vision. We assume that, according to a certain sampling period, features in both image-planes of the stereo system are extracted, stereo-mapped and tracked over successive images of the stereo-image sequence. Some issues related to stereo image processing are delt with in sec. 3.5. The organization of an estimation process that processes noisy stereo-image data is described in sec.3.6. Finally, some results are presented in sec. 3.7.

3.3 Kinematics of relative point-movements

Consider an arbitrary point r_p moving in space with velocity $\dot{r}_p$. The transformation of coordinates between the world- and observer coordinate system is a translation followed by a rotation :

$$\underline{r}'_p = \underline{R}_c (\underline{r}_p - \underline{r}_{o'}) \tag{3-4}$$

where $\underline{r}_{o'}$ denotes the position of the observer origin O', and $\underline{R}_c$ is a rotation matrix.

In order to see how the relative velocities $\dot{r}'_{o''}$ and $\tilde{\underline{\omega}}$ from (3-3) depend on the absolute velocities $\dot{r}_{o'}$, $\dot{r}_{o''}$ $\underline{\omega}_c$ and $\underline{\omega}_b$ we need to know how the velocity of P is expressed in the coordinate frame of the moving observer.

As is shown in appendix A, this is

$$\dot{\underline{r}}'_p = - \underline{\omega}_c \times \underline{r}'_p + \underline{R}_c (\dot{\underline{r}}_p - \dot{\underline{r}}_{o'}) \tag{3-5}$$

Applying this rule to the velocities of the object origin O" and object point P", see fig.3-1, yields

$$\dot{\underline{r}}'_{o''} = - \underline{\omega}_c \times \underline{r}'_{o''} + \underline{R}_c (\dot{\underline{r}}_{o''} - \dot{\underline{r}}_{o'}) \tag{3-6}$$

and

$$\dot{\underline{r}}'_{p''} = - \underline{\omega}_c \times \underline{r}'_{p''} + \underline{R}_c (\dot{\underline{r}}_{p''} - \dot{\underline{r}}_{o'}) \tag{3-7}$$

Substituting (3-1) into (3-7) gives

$$\dot{\underline{r}}'_{p''} = - \underline{\omega}_c \times \underline{r}'_{p''} + \underline{R}_c (\dot{\underline{r}}_{o''} - \dot{\underline{r}}_{o'}) + \underline{R}_c [\underline{\omega}_b \times (\underline{r}_{p''} - \underline{r}_{o''})] \tag{3-8}$$

Since , taking the cross-product of two vectors and rotating the result commute (see appendix A), this , using (3-6), results in

$$\dot{\underline{r}}'_{p''} = - \underline{\omega}_c \times \underline{r}'_{p''} + \dot{\underline{r}}'_{o''} + \underline{\omega}_c \times \underline{r}'_{o''} + \underline{R}_c \, \underline{\omega}_b \times \underline{R}_c (\underline{r}_{p''} - \underline{r}_{o''})] \tag{3-9}$$

where the term $\underline{\omega}'_b := \underline{R}_c \, \underline{\omega}_b$ represents the angular velocity of the object in terms of the observer coordinates and $\underline{R}_c (\underline{r}_{p''} - \underline{r}_{o''})$ may be rewritten, see (3-4)

$$\underline{R}_c (\underline{r}_{p''} - \underline{r}_{o''}) = \underline{R}_c (\underline{r}_{p''} - \underline{r}_{o'}) - \underline{R}_c (\underline{r}_{o''} - \underline{r}_{o'}) = \underline{r}'_{p''} - \underline{r}'_{o''}$$

Substitution into (3-9) results in

$$\dot{r}'_{p''} = \dot{r}'_{o''} + (\underline{\omega}'_b - \underline{\omega}_c) \times (r'_{p''} - r'_{o''}) \tag{3-10}$$

Comparing this relationship with (3-3) we see that the relative angular velocity is the difference of the transformed angular velocity of the object and the angular velocity of the observer. Furthermore, we see that the relative translational velocity given by (3-6) is composed of three components: A rotational component (first term), and two components resulting from the individual movements of the observer- and object-origin. Thus, we formally define the quantities

$$\underline{v}'_{o'} := \underline{R}_c \, \dot{r}_{o'} \quad , \quad \underline{v}'_{o''} := \underline{R}_c \, \dot{r}_{o''}$$

which are the velocities of the observer and object origin resolved in coordinates of the observer. Since $\dot{r}_{o'}$ and $\dot{r}_{o''}$ are free vectors, the time derivatives of their transformed versions $\underline{v}'_{o'}$ and $\underline{v}'_{o''}$ are obtained by applying the rule (3-5) and setting $r_{o'}$ to zero, see (3-4):

$$\dot{\underline{v}}'_{o'} = -\underline{\omega}_c \times \underline{v}'_{o'} + \underline{R}_c \, \ddot{r}_{o'} \tag{3-11}$$

$$\dot{\underline{v}}'_{o''} = -\underline{\omega}_c \times \underline{v}'_{o''} + \underline{R}_c \, \ddot{r}_{o''} \tag{3-12}$$

Similarly, we get for the transformed accelerations $\underline{b}'_{o'} := \underline{R}_c \, \ddot{r}_{o'}$, $\underline{b}'_{o''} := \underline{R}_c \, \ddot{r}_{o''}$ and the time derivative of $\underline{\omega}_b'$

$$\dot{\underline{b}}'_{o'} = -\underline{\omega}_c \times \underline{b}'_{o'} + \underline{R}_c \, \dddot{r}_{o'} \tag{3-13}$$

$$\dot{\underline{b}}'_{o''} = -\underline{\omega}_c \times \underline{b}'_{o''} + \underline{R}_c \, \dddot{r}_{o''} \tag{3-14}$$

$$\dot{\underline{\omega}}'_b = -\underline{\omega}_c \times \underline{\omega}'_b + \underline{R}_c \, \dot{\underline{\omega}}_b \tag{3-15}$$

Finally, we define the quantities

$$\underline{\alpha}'_{o'} := \underline{R}_c \, \dddot{r}_{o'} \quad , \quad \underline{\alpha}'_{o''} := \underline{R}_c \, \dddot{r}_{o''} \quad ,$$

$$\underline{b}'_{\omega b} := \underline{R}_c \, \dot{\underline{\omega}}_b \quad , \quad \underline{b}_{\omega c} := \dot{\underline{\omega}}_c \tag{3-16}$$

With these definitions and writing the cross-products using an antisymmetric matrix $\underline{\Omega}(\underline{\omega})$: $\underline{\omega} \times \underline{r} = \underline{\Omega}(\underline{\omega}) \cdot \underline{r}$, where

$$\underline{\Omega}(\underline{\omega}) = \begin{pmatrix} 0 & -\omega_z & \omega_y \\ \omega_z & 0 & -\omega_x \\ -\omega_y & \omega_x & 0 \end{pmatrix}$$

Equations (3-10), (3-6), and (3-11)-(3-15) may be rewritten:

$$\dot{\underline{r}}'_{p''} = \dot{\underline{r}}'_{o''} + \underline{\Omega}\,(\,\underline{\omega}'_b - \underline{\omega}_c\,) \cdot (\,\underline{r}'_{p''} - \underline{r}'_{o''}\,) \tag{3-17a}$$

$$\dot{\underline{r}}'_{o''} = -\underline{\Omega}\,(\,\underline{\omega}_c\,) \cdot \underline{r}'_{o''} + \underline{v}'_{o''} - \underline{v}'_{o'} \tag{3-17b}$$

$$\dot{\underline{v}}'_{o'} = -\underline{\Omega}\,(\,\underline{\omega}_c\,) \cdot \underline{v}'_{o'} + \underline{b}'_{o'} \tag{3-17c}$$

$$\dot{\underline{v}}'_{o''} = -\underline{\Omega}\,(\,\underline{\omega}_c\,) \cdot \underline{v}'_{o''} + \underline{b}'_{o''} \tag{3-17d}$$

$$\dot{\underline{b}}'_{o'} = -\underline{\Omega}\,(\,\underline{\omega}_c\,) \cdot \underline{b}'_{o'} + \underline{\alpha}'_{o'} \tag{3-17e}$$

$$\dot{\underline{b}}'_{o''} = -\underline{\Omega}\,(\,\underline{\omega}_c\,) \cdot \underline{b}'_{o''} + \underline{\alpha}'_{o''} \tag{3-17f}$$

$$\dot{\underline{\omega}}'_b = -\underline{\Omega}\,(\,\underline{\omega}_c\,) \cdot \underline{\omega}'_b + \underline{b}'_{\omega b} \tag{3-17g}$$

$$\dot{\underline{\omega}}_c = \underline{b}_{\omega c} \tag{3-17h}$$

A few remarks concerning the significance of the quantities $\underline{b}_{\omega c}$, $\underline{b}'_{\omega b}$, $\underline{\alpha}'_{o'} := \underline{R}_c \bar{\underline{r}}_{o'}$, $\underline{\alpha}'_{o''} := \underline{R}_c \bar{\underline{r}}_{o''}$ are in place here. Consider the following special, but important case (We only treat the object movements, similar arguments hold for the observer movements): Let the object rotate at a constant angular speed $\underline{\omega}_b$ around an axis of rotation which is translated with constant acceleration $\ddot{\underline{r}}_z$, $\underline{r}_z$ denoting the position of a point lying on the rotation axis. Clearly, in this case $\underline{b}'_{\omega b}$ is identically zero. In case that the object origin is defined to lie on the rotation axis, say $\underline{r}_{o''} = \underline{r}_z$, then also $\bar{\underline{r}}_{o''}$ and therefore $\underline{\alpha}'_{o''}$ are identically zero. Often however, such a definition is not reasonable, for example, when the axis of rotation is not known or located outside the object. Then similar to eq. (3-1) we have

$$\dot{\underline{r}}_{o''} = \dot{\underline{r}}_z + \underline{\omega}_b \times (\,\underline{r}_{o''} - \underline{r}_z\,) = \dot{\underline{r}}_z + \underline{\Omega}(\underline{\omega}_b)\,(\,\underline{r}_{o''} - \underline{r}_z\,)$$

$$\Rightarrow \quad \ddot{\underline{r}}_{o''} = \ddot{\underline{r}}_z + \underline{\Omega}(\dot{\underline{\omega}}_b)\,(\,\underline{r}_{o''} - \underline{r}_z\,) + \underline{\Omega}(\underline{\omega}_b)\,(\,\dot{\underline{r}}_{o''} - \dot{\underline{r}}_z\,)$$

$$= \ddot{\underline{r}}_z + \underline{\Omega}^2(\underline{\omega}_b)\,(\,\underline{r}_{o''} - \underline{r}_z\,) \qquad (\,since\ \dot{\underline{\omega}}_b \equiv \underline{0}\,)$$

$$\Rightarrow \quad \bar{\underline{r}}_{o''} = \bar{\underline{r}}_z + \underline{\Omega}^2\,(\underline{\omega}_b)\,(\,\dot{\underline{r}}_{o''} - \dot{\underline{r}}_z\,)$$

$$= \underline{\Omega}^3(\underline{\omega}_b)\,(\,\underline{r}_{o''} - \underline{r}_z\,) \qquad (\,since\ \bar{\underline{r}}_z \equiv \underline{0}\,)$$

$$= \underline{\omega}_b \times \{\,\underline{\omega}_b \times [\,\underline{\omega}_b \times (\,\underline{r}_{o''} - \underline{r}_z\,)^+\,]\,\}$$

where $(\underline{r}_{o''} - \underline{r}_z)^+$ denotes the component of the difference vector $(\underline{r}_{o''} - \underline{r}_z)$ which is orthogonal to $\underline{\omega}_b$. Hence, it follows

$$|| \bar{r}_{o''} || = || \underline{\omega}_b |^{\beta} \cdot || (\underline{r}_{o''} - \underline{r}_z)^+ ||$$

Thus, the magnitude of $\underline{\alpha}'_{o''}$ (which is the same as that of $\bar{r}_{o''}$) is small in case $|| \underline{\omega}_b ||$ is small and/or the distance of the object origin from the rotation axis is small. Then the higher order derivatives $\underline{\alpha}'_{o''}$ may be treated as small disturbances and, therefore, no further modeling (differential equations) is needed. We return to this point in sec. 3.6.

3.4 Representation of relative pose and object-structure

Relative pose

Again, consider an arbitrary point P" fixed on the object with position vector in object-centered coordinates $\underline{r}''_{P''}$. Its representation w.r.t. the observer frame is obtained by translation and rotation according to

$$\underline{r}'_{P''} = \underline{r}'_{o''} + \tilde{\underline{R}} \cdot \underline{r}''_{P''} \tag{3-18}$$

Here, $\underline{r}'_{o''}$ represents the *relative position* and the rotation matrix $\tilde{\underline{R}}$ specifies the *relative orientation* of the object w.r.t. the observer. Both, relative position and relative orientation, specify the *relative pose* of the object.

Differentiating (3-18) , keeping in mind that $\dot{\underline{r}}''_{P''} \equiv \underline{0}$ and comparing the result with (3-17a) yields:

$$\dot{\tilde{\underline{R}}} = \underline{\Omega} (\underline{\omega}'_b - \underline{\omega}_c) \cdot \tilde{\underline{R}} \tag{3-19}$$

Thus, we have differential equations for the rotation matrix $\tilde{\underline{R}}$ describing the change of relative orientation. Representing orientations by rotation matrices requires the specification of nine parameters, although it is well known that any rotation matrix is a function of only three independent parameters, for example Euler angles or roll-pitch-yaw angles, [PAUL 81]. In case a representation by roll-pitch-yaw angles $\underline{\varphi}^T = [\alpha, \beta, \gamma]$ is chosen (denoting successive rotations about the x-, y- and z- axis respectively), this relation becomes ($\sin\alpha := s\alpha$, $\cos\alpha := c\alpha$, etc.)

$$\tilde{\underline{R}} (\underline{\varphi}) = \begin{pmatrix} c\gamma\, c\beta & -s\gamma\, c\alpha + c\gamma\, s\beta\, s\alpha & s\gamma\, s\alpha + c\gamma\, s\beta\, c\alpha \\ s\gamma\, c\beta & c\gamma\, c\alpha + s\gamma\, s\beta\, s\alpha & -c\gamma\, s\alpha + s\gamma\, s\beta\, c\alpha \\ -s\beta & c\beta\, s\alpha & c\beta\, c\alpha \end{pmatrix}$$

Upon calculating $\dot{\tilde{\underline{R}}}(\underline{\varphi})$ and substituting into (3-19), one finds a relation between the time derivatives of $\underline{\varphi}$ and the relative angular velocity $\tilde{\underline{\omega}} = \underline{\omega}'_b - \underline{\omega}_c$:

$$\tilde{\underline{\omega}} = \underline{C}(\underline{\varphi}) \cdot \dot{\underline{\varphi}} \quad , \quad \dot{\underline{\varphi}} = \underline{C}^{-1}(\underline{\varphi}) \cdot \tilde{\underline{\omega}} \tag{3-20}$$

where

$$\underline{C}(\underline{\varphi}) = \begin{pmatrix} c\gamma\, c\beta & -s\gamma & 0 \\ s\gamma\, c\beta & c\gamma & 0 \\ -s\beta & 0 & 1 \end{pmatrix}$$

However, for $\beta = (2k+1)\,\pi/2$, $k = ...-1, 0, 1, ...$, $\underline{C}(\underline{\varphi})$ is singular and therefore the second relation is undefined. It turns out, that any representation involving only three parameters has singular points in this sense , [STUELPNAGEL 64].

For this reason a representation by unit quaternions, [HORN 87] is chosen which is nonsingular (see also [SPRING 86] for a comparison of the most common representations of orientations). Given the unit vector $\underline{n}$ of the rotation axis and θ the rotation angle, the unit quaternion representing $\tilde{\underline{R}}$ is defined to be

$$\underline{\kappa}^T = [\sin(\frac{\theta}{2}) \cdot \underline{n}^T, \cos(\frac{\theta}{2})] := [\underline{\kappa}_v^T, \kappa_s] \tag{3-21}$$

where $\underline{\kappa}_v$ is the so called "vector part" of the quaternion and κ_s its "scalar part". Some details of the quaternion-representation of rotations are given in appendix B. There it is shown that an analogous relation to (3-20) relating the angular velocity $\tilde{\underline{\omega}}$ and the time derivative of the unit quaternion representing $\underline{R}$ is given by

$$\dot{\underline{\kappa}} = \frac{1}{2} \cdot \underline{\Gamma}(\tilde{\underline{\omega}}) \cdot \underline{\kappa} \tag{3-22}$$

where

$$\underline{\Gamma}(\tilde{\underline{\omega}}) = \begin{pmatrix} 0 & \tilde{\omega}_z & -\tilde{\omega}_y & \tilde{\omega}_x \\ -\tilde{\omega}_z & 0 & \tilde{\omega}_x & \tilde{\omega}_y \\ \tilde{\omega}_y & -\tilde{\omega}_x & 0 & \tilde{\omega}_z \\ -\tilde{\omega}_x & -\tilde{\omega}_y & -\tilde{\omega}_z & 0 \end{pmatrix}$$

A slight drawback of the representation by quaternions is that it is not unique. As can by seen in appendix B $\underline{\kappa}$ and $-\underline{\kappa}$ represent the same rotation. Thus, by selecting one of them, uniqueness can be achieved.

In the sequel we assume that the rotation angle θ satisfies $-\pi < \theta < \pi$.

In this case, the representation is clearly unique since the sign of $\cos(\theta/2)$ is always positive.

Although the dynamics of rotations (eq. (3-22)) requires the vector and scalar part of the unit quaternion, the vector part suffices to specify the rotation $\tilde{R}$, since it defines $\underline{n}$ and θ up to a sign ambiguity (which is not important since $-\underline{n}, -\theta$ define the same rotation).

Summing up, the relative pose may be represented by a 6-dimensional vector $\underline{s}'^T = [\underline{r}'^T_{o''} \; \underline{\kappa}^T_v]$.

Object - structure

As cited in the introduction, the object structure is characterized by a set of prominent points on the object of interest.

The relation (3-18) suggests two possibilities:

1. representation in an observer-centered coordinate frame, i.e. by means of a set of position vectors $\underline{r}'_i$, $i = 1$, ... , M or equivalently by difference-vectors $\underline{r}'_i - \underline{r}'_{o''}$ that are rigidly connected with the object

2. representation in an object-centered coordinate frame, i.e. by means of a set of position vectors $\underline{r}''_i$, $i = 1$, ... , M.

It has often been claimed, that an object centered representation is preferable, since the geometrical structure is an invariant property of a rigid object, that should be separated from movements of an observer. Therefore, we choose the object-centered representation and call the set of M prominent points ($\underline{r}''_i$, $i = 1$, $\cdots$, M) the "structure parameters".

3.5 Processing stereo images

It is well known that the position of a point P in 3D-space can be infered from its projections into two image-planes which are geometrically related to each other in a known way, called a "stereo setup" (or "stereo pair").

We consider a general stereo setup as depicted in fig. 3-2.

To each camera of the stereo setup an image coordinate frame (C_l, x_l, z_l) , (C_r, x_r, z_r) , for the left and right camera respectively) and a 3D-coordinate frame ($O_l, x_{kl}, y_{kl}, z_{kl}$) , ($O_r, x_{kr}, y_{kr}, z_{kr}$) is associated, the y_{kl} -and y_{kr} axis being orthogonal to the respective image planes; (O', x', y', z') as before denotes the observer coordinate frame.

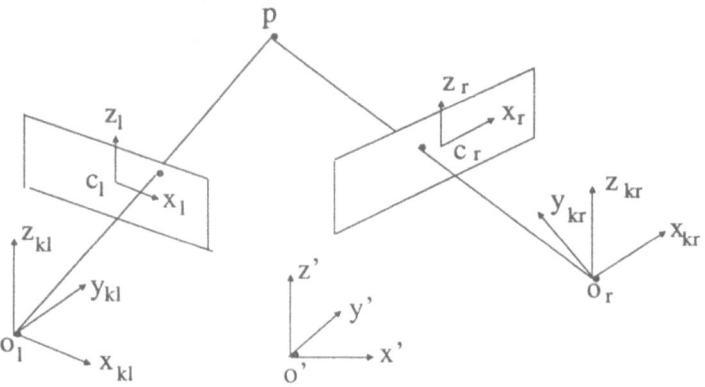

Fig. 3-2 Stereo setup in a general configuration

.According to a simple pin-hole camera model, the physical image coordinates of the projections of a point P in space are related to its 3D-camera coordinates by the laws of perspective transformations, [DUDA, HART 73]

$$\frac{x_l}{f_l} = \frac{x_{kl}}{y_{kl}} \qquad , \qquad \frac{x_r}{f_r} = \frac{x_{kr}}{y_{kr}}$$

$$\frac{z_l}{f_l} = \frac{z_{kl}}{y_{kl}} \qquad , \qquad \frac{z_r}{f_r} = \frac{z_{kr}}{y_{kr}} \qquad (3\text{-}24)$$

where f_l and f_r denote the focal lengths of the cameras.

The 3D camera frames are related to the observer frame by rotations and translations according to

$$\underline{r}_{kl} = \underline{R}_l \cdot \underline{r}' + \underline{T}_l \qquad , \qquad \underline{r}_{kr} = \underline{R}_r \cdot \underline{r}' + \underline{T}_r \qquad (3\text{-}25)$$

Writing the rotation matrices and translation vectors in the form

$$\underline{R}_l = \begin{pmatrix} \underline{e}_{1l}^T \\ \underline{e}_{2l}^T \\ \underline{e}_{3l}^T \end{pmatrix} \qquad ; \qquad \underline{R}_r = \begin{pmatrix} \underline{e}_{1r}^T \\ \underline{e}_{2r}^T \\ \underline{e}_{3r}^T \end{pmatrix}$$

$$\underline{T}_l = \begin{pmatrix} t_{xl} \\ t_{yl} \\ t_{zl} \end{pmatrix} \qquad ; \qquad \underline{T}_r = \begin{pmatrix} t_{xr} \\ t_{yr} \\ t_{zr} \end{pmatrix}$$

and combining (3-24) and (3-25), relations for the image- and 3D-observer coordinates of P are obtained

$$\frac{x_l}{f_l} = \frac{\underline{e}_{1l}^T \cdot \underline{r}' + t_{xl}}{\underline{e}_{2l}^T \cdot \underline{r}' + t_{yl}} \quad ; \quad \frac{x_r}{f_r} = \frac{\underline{e}_{1r}^T \cdot \underline{r}' + t_{xr}}{\underline{e}_{2r}^T \cdot \underline{r}' + t_{yr}}$$

$$\frac{z_l}{f_l} = \frac{\underline{e}_{3l}^T \cdot \underline{r}' + t_{zl}}{\underline{e}_{2l}^T \cdot \underline{r}' + t_{yl}} \quad ; \quad \frac{z_r}{f_r} = \frac{\underline{e}_{3r}^T \cdot \underline{r}' + t_{zr}}{\underline{e}_{2r}^T \cdot \underline{r}' + t_{yr}} \qquad (3\text{-}26)$$

In case that the geometrical parameters of the stereo setup are known, and given the projections of a point P in 3D-space, its coordinates w.r.t. the observer coordinate frame can be calculated by solving the relationships (3-26) (However, as will become clear in sec. 3.6 dealing with the estimation process, direct inversions of (3-26) are not needed).

Thus, inferring the 3D-position of a point P involves finding its corresponding images in the two views of the stereo pair. Furthermore, in case of dynamic scenes, the image points of P have to be tracked over time. These issues are discussed in the following two paragraphs.

Stereo-matching of image descriptions

This step is the most difficult in stereo image processing. The problem is: Given a point of interest in one image, what is then the location of the, to the same 3D-point corresponding point in the other image ? Image points that are to be matched in this sense, are often called "homologous". This definition similarly applies to other image entities, such as straight line segments.

Since single image points (pixels) carry only little information (the greyvalue), they do not provide enough distinguishing power for unambigous point to point matching. Practical algorithms rely either on similarity of small areas centered around the point of interest or on matching prominent parts of an image (features), such as edges, grey-value corners, blobs etc.

Approaches of the first category often apply image intensity correlation techniques, [MORI et al. 73], [YAKIMOVSKY, CUNNINGHAM 78], [MORAVEC 79], [HOBROUGH, HOBROUGH 83].

It is well known that matching two similar signals by maximizing their correlation is optimal if the difference of the two signals is white noise [VAN TREES 68]. However, in the case of stereo-matching this assumption is often violated, since the grey value function of the right image is a distorted version of that in the left image due to different viewing angles. This distortion is in general not adequately modeled by white noise. Furthermore, problems arising from repetitive structures puzzle any scheme relying on correlation.

The second category of matching algorithms mentioned above , is often called "feature based stereo". Features (mostly edges) are extracted from grey-level images and matched according to certain strategies which cope with the combinatorical explosion of possible pairings and avoid false matchings (often called "false targets") [MARR, POGGIO 79], [GRIMSON 81], [NISHIHARA 83], [POLLARD et al. 87], [AYACHE, FAVERJON 87].

We adopted the concept underlying the approach of Ayache and Faverjon which relies on a symbolic description of images using line segments that are matched according to a hypothesize and test strategy. Based on this concept we developed an algorithm showing a strong overall resemblance to their algorithm but differing substantially in a number of points, leading to enhanced performance under certain circumstances, [PUSCH 90].

This concept will now be briefly described.

1. step: Feature extraction

Starting from the grey-level images of the left and right camera, edge detection is performed on each image. We used Haralicks facet-model [HARALICK 84] for locally expanding the grey-level function by orthogonal polynoms. At each pixel location, the gradient of the gray-level function is estimated from the coefficients of these expansions. After an adaptive thresholding performed according to certain image statistics, the second directional derivative in the direction of the gradient is computed. The zero crossings of the second directional derivative are marked as edge points resulting in a binary edge image. Although the adaptive thresholding removes "false" edges due to noise, some spurious edges remain because smoothing is kept to a minimum in the edge detection process, in order to retain a high level of detail and avoid rounding of corners etc.

2. step: Segmentation

The edge-images are segmented by approximation of edge-contours with linear segments, using a very efficient scan-along technique developed by [WALL, DANIELLSON 84].

After deleting very short segments considered unreliable and often due to spurious edges, a symbolic description of the images is made, consisting of two lists of straight line segments with, attached to each line segment, a list of properties:

- identification

- position of its endpoints

- orientation

- length

3. step: Matching of line segments

The third step involves the actual matching of line segments from the left image to those of the right image. This is done using certain constraints for the reduction of the number of possible pairings:

1. uniqueness constraint: For each line segment in the left image there is at most one corresponding line segment in the right image.

2. epipolar constraint: Consider fig.3-2. Given a feature point in the left image, the corresponding 3D-point P may lie everywhere on the straight line joining P and the left focal center O_l (called "projecting ray"). Therefore, the homologous image point in the right image is located on the projection of the projecting ray into the right image plane; a straight line called the "epipolar line". In this way, the search space is reduced to the epipolar line, which may be computed from the feature position in the left image and the parameters of the stereo setup. The epipolar constraint can also be used for matching straight line segments if it is applied to their endpoints.

3. 3D- continuity constraint: The 3D-coordinates of a point P sliding on the surface of an object changes continously everywhere . Since the relations (3-26) define a continous mapping $\underline{r}' \rightarrow (x_l, z_l, x_r, z_r)$, the image coordinates of the projections of P change continuously too. Given a homologous pair P_l, P_r and another point P_l' near to P_l, then it is likely that its homologous counterpart P_r' is near to P_r since the corresponding 3D-points P and P' are likely to be located on a surface of the same object. This fact may be taken into account during the matching process as follows: Consider the special stereo setup as depicted in fig. 3-3 (the general configuration of fig. 3-2 can be treated similarly).

As can be shown by evaluating (3-26), the depth of two neighboring 3D-points P_1, P_2 with coordinates $\underline{r}'$ and $\underline{r}' + d\underline{r}'$ may be calculated, given their projections, as (for simplicity we assume equal focal lengths: $f_l = f_r =: f$)

$$y_1' = \frac{D}{\kappa_{l1} - \kappa_{r1}} \qquad , \qquad y_2' = \frac{D}{\kappa_{l2} - \kappa_{r2}} \tag{3-27}$$

where

$$\kappa_{li} := \frac{x_{li}/f + \tan \varphi}{1 - \tan \varphi \cdot x_{li}/f}$$

$$\kappa_{ri} := \frac{x_{ri}/f - \tan \varphi}{1 + \tan \varphi \cdot x_{ri}/f}$$

We call $v_i := \kappa_{li} - \kappa_{ri}$ the "disparity" belonging to P_i. The depth of P_2 as a function of its disparity can be expanded into a Taylor series around the depth of P_1 :

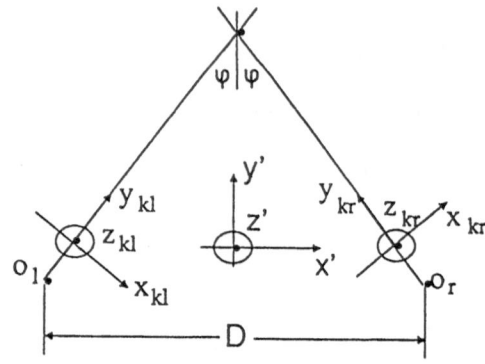

Fig. 3-3   Stereo setup with equally verged cameras

$$y'(v_1 + dv) = y'(v_1) + \frac{\partial}{\partial v} y'(v) \big|_{v = v1} dv + O (dv^2)$$

$$\approx y'_1(v_1) - \frac{D}{v_1^2} dv \tag{3-28}$$

If the magnitude of the difference of depths $dy'(v_1) := y(v_1 + dv) - y(v_1)$ is assumed to be bounded by some constant $\varepsilon > 0$, i.e. $| dy'(v_1)| \geq \varepsilon$ then the magnitude of the difference of disparities is limited by a constant δ according to

$$| dv| \leq \delta = \frac{\varepsilon v_1^2}{D} \tag{3-29}$$

Given an $\varepsilon > 0$, (3-29) poses a constraint on the disparities of two neighboring image points. Note, that for calculating δ only the x-coordinates of the candidate matches and the geometric parameters of the stereo setup are needed. The definition of disparity as given above is also suitable for straight line segments if it is applied to the endpoints of a segment.

4. local constraints: Assuming that the geometry of the stereo setup renders the viewing angles of the two cameras not too different, the orientations and lengths of homologous straight line segments are likely to be approximately the same.

Stereo-matching by 3D-line segment chaining

Now, the strategy of our stereo-matching algorithm which uses the constraints defined above can be described as follows.

After the segmentation process is completed, two lists of straight line segments, one for each image of the stereo setup, are available. Consider one segment in the left image denoted by S_{li} see fig.3-4.

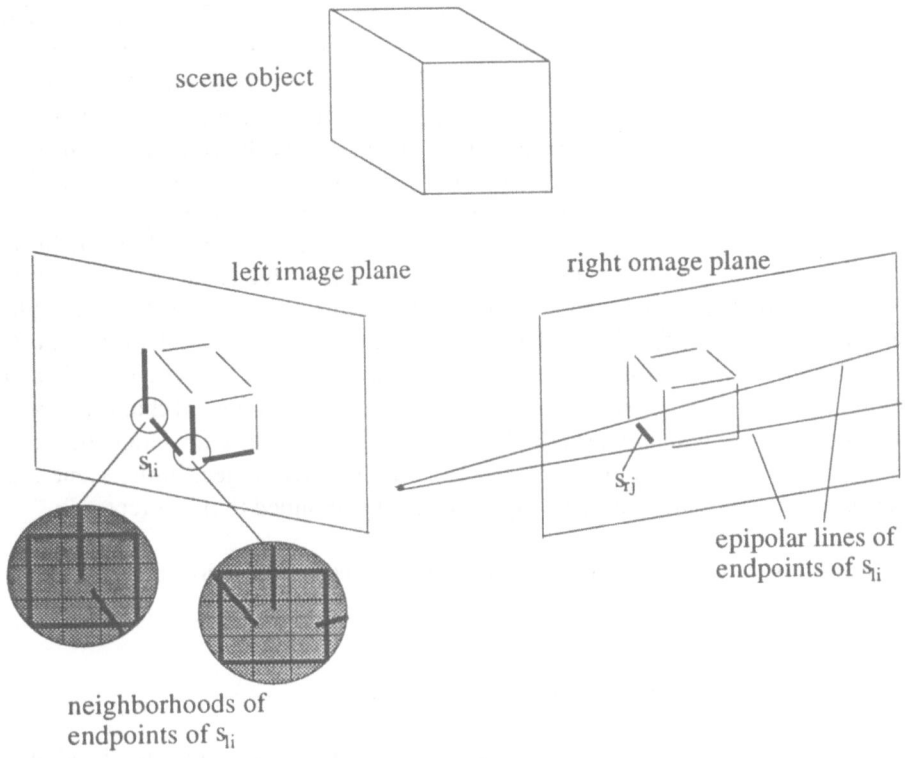

Fig. 3-4 Chaining of line segments

From the list of segments in the right image which satisfy the epipolar and local constraints the one that best fits to S_{li}, denoted by S_{rj}, is selected and the disparities of its endpoints, denoted by $v_1(i,j)$ and $v_2(i,j)$ respectively, calculated. The pair (S_{li}, S_{rj}) defines a hypothetical 3D-straight line segment which is called "starting hypothesis". Assuming this 3D-segment to be derived from a real object in space, it is likely that there are other 3D-segments connected to the same object. Or, the other way round, if other 3D-segments can be found that are located in the neighborhood of the starting segment, the likeliness that the starting hypothesis is correct increases. In order to arrive at a functioning algorithm, it is crucial to properly define the "neighborhood of 3D-line segments". We define two 3D-segments to be neighbors iff they have any neighboring endpoints (see fig.3-4). The search for neighboring endpoints may be done in the image planes. According to a predefined partition of the image planes into small squares, a local neighborhood centered around one of the endpoints of S_{li} is defined, and it is checked if there are any endpoints of other segments in this neighborhood. If this is the case for some S_{ln}, then it is checked whether there is a fitting right-segment that satisfies the epipolar, local and continuity constraint (with suitable ε). In this case, S_{ln} is connected to S_{li}. This process is called "chaining". The same procedure is done for all neighbors of S_{li} until no further hypothetical matches can be found. Hence, beginning with the starting hypothesis, a "chain" of connected 2D-segments in the left image is defined that corresponds to a chain of 2D-segments in the right image. Since all left-right matches contained in these chains satisfy the continuity constraint, it is likely that the corresponding hypothetical 3D-segments defined by each matching pair belong to the same continous region in space, i.e. to the same object. However, since the whole chaining grounds on the starting hypothesis, the matching chains are itself hypothetical. Moreover, the chaining is attempted for each left-segment and therefore it is likely that there are segments for which more than one match is assigned, say (S_{li}, S_{rj}) and (S_{li}, S_{rk}). The case $j = k$ causes no problem, however if $j \neq k$ a conflict with the uniqueness constraint occurs. Let C be the set of all chains containing (S_{li}, S_{rk}). Then the conflicts are resolved by deleting all chains contained in C except the one with highest length. This is reasonable, since the length of a chain is a measure for likeliness of the chain belonging to a real object in the scene , and hence the matches contained in the largest chain are assumed to be correct.

In appendix C a brief description of an efficient implementation of the proposed strategy is given.

Matching of line segments in time

Since relative movements and object structure are to be estimated using visual meas- urements from successive images of a stereo sequence, image features (here: straight line segments) have to be tracked over time. This poses a further matching problem which may be solved in a similar way as the stereo matching problem. All constraints described above may be used to reduce the number of potential matches, except the epipolar constraint, since the relative motion induced by the individual motions of the observer and object are arbitrary. However, the matching of features in time is in principle easy compared to the stereo matching problem, provided that the sampling interval of the stereo sequence is short enough compared to the velocities of relative movements. In this case, the locations of visual features in the image planes of the stereo setup do not change significantly between two successive images and therefore unambigous matches may be easily found. Furthermore, the tracking of features is alleviated by exploiting information from predictions for the relative pose of the object of interest, which are provided by the estimation filters to be discussed in sec. 3.6 at every sampling instant. Hence, the locations of features at time t_{k+1} may be predicted by applying the laws of perspective transformations (3-26).

These issues are currently under investigation and will not be further discussed here.

Measurements for structure and relative position

Given a pair of homologous image points P_l , P_r the 3D observer coordinates of the corresponding point P in space may be calculated using the equations (3-26). Thus, having homologous projections for the M 3D-points defining object structure and the origin of the object coordinate frame O", (3-26) constitute 4(M+1) measurements which can be used for structure and relative movement estimation. Since the output of the stereo matcher is a list of homologous straight line segments, prominent image points have to be defined in terms of these. Among many others, the following possibilities for defining homologous points exist:

1. by intersections of homologous straight lines (e.g. vertices of polyhedra), or

2. by endpoints of homologous straight line segments, or

3. by fixing a point P_l on a straight line segment in the left image and getting its homologous counterpart P_r in the right image by finding the intersection of the homologous right segment with the epipolar line belonging to P_l .

Measurements for the relative orientation

In order to derive estimates for the relative orientation $\overline{\kappa}$ from visual measurements, some a priori information about the object being observed is needed. This means that we must require, that certain visual features of the object, being uniquely related to the object-centered coordinate frame, can be identified during the estimation-process.

Many authors who address the problem of orientation estimation assume that a priori knowledge about the positions of a set of structure-points is available, see for example [YOUNG, CHELLAPPA 88], [WU et al. 88]. This assumption severely limits the flexibility of these approaches, since these techniques fail if the size of the observed object differs from that of the model.
In fact, since the relative object-orientation is identical with the orientation of the object coordinate frame w.r.t. the observer frame, it suffices to require, that two of the basis vectors of the object frame can be recovered from stereoscopic images during the estimation process. More generally, we require a priori knowledge of N non-colinear "intrinsic prominent directions" of the object which are represented by unit vectors with coordinates w.r.t. the object frame:

$$\underline{d_i}" \; : \; i = 1, \cdots, N \quad , \quad N \geq 2 \quad , \quad || \underline{d_i}" || \; = 1 \quad \text{for } all \; i \quad (3\text{-}32)$$

The intrinsic prominent directions d_i" transform into directions seen by the observer according to the time-dependent relative orientation represented by a unit quaternion $\bar{\kappa}(t)$:

$$d_i'(t) = \tilde{R} \cdot d_i" = \tilde{R}(\bar{\kappa}(t)) \cdot d_i" \tag{3-33}$$

where $\tilde{R}(\bar{\kappa}(t))$ is given in appendix B . The transformed directions d_i' are simply called "prominent directions" in the sequel to distinguish them from the intrinsic prominent directions $d_i"$.

Assuming that we have measurements for the prominent directions d_i' , the relations (3-33) constitute 3N nonlinear measurement equations for $\bar{\kappa}$. Thus, it remains to be specified, how measurements for d_i' may be extracted from stereoscopic images.

A simple possibility would be to take difference-vectors $r_i' - r_j'$, $i \neq j$, defined by structure points. r_i' , r_j' can be calculated by inversion of eq. (3-26). Assuming, that the directions of $r_i' - r_j'$ are known a priori, they could be taken as prominent directions. It can be shown, that the errors in these measurements increase quadratically with the object's distance from the observer. Another disadvantage of this approach is that it depends on the presence of structure points being localizable in the two images of the stereo system which may not always be the case. For these reasons, the measurement of prominent directions will be based on the concept of straight lines.

Conceptually, we may associate a 3D-straight line, given by the parametric representation

$$r' = r_g' + \lambda d_i' \tag{3-34}$$

to each prominent direction d_i', where r_g' is a position vector pointing to an arbitrary point of the straight line and λ is a real parameter. Each point r' of the straight line maps to the image planes of the stereo system according to the laws of perspective transformations, see eq. (3-26) (for brevity, we only consider the left image and as before assume equal focal lengths) :

$$x_l = f \frac{e_{1l}^T \cdot r' + t_{xl}}{e_{2l}^T \cdot r' + t_{yl}} \quad ; \quad z_l = f \frac{e_{3l}^T \cdot r' + t_{zl}}{e_{2l}^T \cdot r' + t_{yl}} \tag{3-35}$$

Since 3D-straight lines project into 2D-straight lines, the image points (x_l, z_l) of (3-35) satisfy the relation

$$z_l = a_l x_l + b_l \tag{3-36a}$$

for lines that are not parallel to the z-axis, or

$$x_l = \tilde{a}_l z_l + \tilde{b}_l \tag{3-36b}$$

for lines that are not parallel to the x-axis.

In the following, we consider the case (3-36a). Substituting (3-34),(3-35) into (3-36a) and multiplying by the denominator gives

$$f \underline{e}_{3l}^T (\underline{r_g}' + \lambda \underline{d_i}') + f t_{zl} = a_l f \underline{e}_{1l}^T (\underline{r_g}' + \lambda \underline{d_i}') + a_l f t_{xl} +$$

$$+ b_l \underline{e}_{2l}^T (\underline{r_g}' + \lambda \underline{d_i}') + b_l t_{yl} \tag{3-37}$$

for all $\lambda \in R$.

Setting $\lambda = 0$ yields

$$f \underline{e}_{3l}^T \underline{r_g}' + f t_{zl} = a_l f \underline{e}_{1l}^T \underline{r_g}' + a_l f t_{xl} + b_l \underline{e}_{2l}^T \underline{r_g}' + b_l t_{yl} \tag{3-38}$$

Setting $\lambda = 1$ in (3-37) and subtracting (3-38) gives

$$f \underline{e}_{3l}^T \underline{d_i}' = a_l f \underline{e}_{1l}^T \underline{d_i}' + b_l \underline{e}_{2l}^T \underline{d_i}' \tag{3-39a}$$

which may be rewritten

$$\underline{d_i}'^T [\underline{e}_{1l} , \underline{e}_{2l} , \underline{e}_{3l}] \begin{pmatrix} f & 0 & 0 \\ 0 & 1 & 0 \\ 0 & 0 & -f \end{pmatrix} \begin{pmatrix} a_l \\ b_l \\ 1 \end{pmatrix} = 0$$

or

$$\underline{d_i}'^T \underline{S_l} \, g_l = 0$$

where

$$\underline{S_l} := \begin{pmatrix} f & 0 & 0 \\ 0 & 1 & 0 \\ 0 & 0 & -f \end{pmatrix} \underline{R_l}^T := \underline{S_o} \underline{R_l}^T$$

$$g_l^T := [a_l , b_l , 1]$$

since $\underline{R_l}^T \underline{S_o} = \underline{S_o} \underline{R_l}^T$.

Similarly, for the representation (3-36b), we have

$$\underline{d_i}'^T \underline{S_l} \, \tilde{g}_l \qquad where \qquad \tilde{g}_l = [1 , -\tilde{b} , \tilde{a}] \tag{3-39b}$$

Now, for the right camera of the stereo system we get a relation similar to (3-39a) resp. (3-39b)

$$\underline{d_i}'^T \underline{S_r} g_r = 0 \qquad\qquad (3\text{-}39c)$$

where

$$\underline{S_r} := \underline{S_o} \underline{R_r^T} \quad , \quad g_r^T := [\, a_r\,, b_r\,, 1\,]$$

or

$$\underline{d_i}'^T \underline{S_r} \tilde{g}_r = 0 \quad , \quad \tilde{g}_r^T := [\, 1\,, -\tilde{b}_r\,, \tilde{a}_r\,] \qquad\qquad (3\text{-}39d)$$

From (3-39a) and (3-39c) it follows, that $\underline{d_i}'$ is orthogonal to $\underline{S_l} g_l$ and $\underline{S_r} g_r$. Therefore, assuming that $\underline{S_l} g_l \neq \underline{S_r} g_r$, the cross product of $\underline{S_l} g_l$ and $\underline{S_r} g_r$ is a vector being collinear with $\underline{d_i}'$:

$$\underline{\mu_i} := \underline{S_l} g_l \times \underline{S_r} g_r = c_i \underline{d_i}' \quad , \quad |c_i| = \left| \underline{S_l} g_l \times \underline{S_r} g_r \right| \qquad\qquad (3\text{-}40)$$

One can show, that the assumption $\underline{S_l} g_l \neq \underline{S_r} g_r$ is violated iff the 3D-straight line is contained in an epipolar plane of the stereo system (a plane containing the two focal centers of the cameras). In these cases the calculation of difference vectors defined by 3D-structure points already mentioned above can be an alternative way for getting measurements for prominent directions.

The matrices $\underline{S_l}$ and $\underline{S_r}$ contain geometric parameters of the stereo system and are therefore available by calibration. Note, that the translation-parameters $\underline{T_l}$ and $\underline{T_r}$ are not required and therefore the measurements of orientation are not corrupted by errors in these parameters.

In many cases the prominent directions $\underline{d_i}'$ respectively $\underline{d_i}''$ may be identified with 3D-straight line segments associated with edges of the object of interest (e.g. in case of a polyhedron). Then the projections of these 3D-straight line segments are visible in the images of the stereo setup. The techniques proposed in sec. 3.4 may be applied directly, i.e. after edge detection and segmentation homologous 2D-line segments are provided by the stereo matcher and the prominent directions calculated according to (3-40). In case a high accuracy is desired, the required 2D straight line parameters can be calculated by applying a least squares fitting technique to the edge images.

In other cases prominent directions can be defined to be functions of other directions calculated by (3-40). For instance, the vector cross product of two directions contained in a plane can be used to define a prominent direction being normal to this plane.

The sign of c_i in (3-40), defining the orientation of a prominent direction $\underline{d_i}'$, must be known. This means that we require from the object recognition process rough knowledge about the viewing direction for relating the intrinsic prominent directions $\underline{d_i}''$, contained in an object-model base, to visual features defining the $\underline{d_i}'$. A rough recognition of the viewing direction must be provided by any object recognition system except for some simple cases, where the viewing direction is known a priori.

3.6 Estimation process

In this section we adress the problem of deriving estimates for the structure and relative motion parameters using stereoscopic measurements being derived from successive image pairs of a stereo-sequence.

In the sequel, we shall denote by t_k the time instants when the coordinates of homologous image points and prominent directions become available for the estimation process.

Since the visual measurements are corrupted by errors rather stochastic in nature, an extended Kalman filtering (EKF) approach is appropriate, [MAYBECK 82].

In order to define filter structures, the dynamics of the involved processes have to be formulated in terms of state space models. Concerning the evolution of the relative pose, eq. (3-17b)-(3-17f) define state space equations for the relative position and the translational motion components and eq. (3-22), (3-17g), (3-17h) account for the relative orientation and rotational motion components. The object structure is represented by the structure parameters $\underline{r}''_i$, $i = 1, \dots , M$ being constant in time. Hence, we have

$$\underline{\dot{r}}''_i \equiv \underline{0} \tag{3-41}$$

The ensemble of state space equations may be written more compactly, (suppressing the time arguments)

$$\underline{\dot{x}} = \underline{f}(\underline{x}, \underline{u}) \tag{3-42}$$

where the state $\underline{x}$ is

$$\underline{x}^T = [\, \underline{r}'^T_{o''}\ \underline{v}'^T_{o'}\, , \underline{v}'^T_{o''}\, , \underline{b}'^T_{o'}\, , \underline{b}'^T_{o''}\, , \overline{\underline{\kappa}}^T\, , \underline{\omega}^T_c\, , \underline{\omega}'^T_b\, , \underline{r}''^T_1\, , \dots , \underline{r}''^T_M\,]$$

and $\underline{u}$ denotes inputs either known or unknown

$$\underline{u}^T = [\, \underline{\alpha}'^T_{o'}\, , \underline{\alpha}'^T_{o''}\, , \underline{b}^T_{\omega c}\, , \underline{b}'^T_{\omega b}\,] \tag{3-42a}$$

Observability-considerations

In order to set up an estimation problem that has a well defined solution, we have to ensure that the state-space model defining the process considered with associated measurements is observable (for the definition of observability, see for example [BOECKER et al. 86]. To be more specific, consider the subset of equations describing the translational motions (3-17b)-(3-17f).

We assume, that the images of the object-origin O" are known. Then , by (3-26), these measurements carry the full information about the relative position $\underline{r}'_{o''}$.

Unfortunately, these measurements are not sufficient for reconstruction of the translational motion states, because the dynamical system (3-17b)-(3-17f) with measurements (3-26) is not observable in general. Since only the difference of the velocities $\underline{v}'_{o'}(t)$, $\underline{v}'_{o''}(t)$ influences the relative position $\underline{r}'_{o''}(t)$, see (3-17b), and therefore the measurements, it is clear that only the difference of the velocities may be recovered from the measurements. Furthermore, since the rotational motions are decoupled from the translational motions, and the structure of the object does not depend on $\underline{v}_{o'}'(t)$ and $\underline{v}_{o''}'(t)$ at all, this problem cannot be overcome by adding more point-measurements or measurements for prominent directions.

In order to ensure observability, we require that at least one of the velocities $\underline{v}_{o'}'(t)$ and $\underline{v}_{o''}'(t)$ is known or being measured.

In case $\underline{v}_{o'}'(t)$ is known with sufficient accuracy, the differential equations (3-17c) and (3-17e) are removed and $\underline{v}_{o'}'(t)$ is treated as a known input. This may be true for example for manipulator control using eyes in hand. Here, $\underline{v}'_{o'}(t)$ is a known function of known joint-coordinates and joint-velocities, see chapter 4.

In an other case, $\underline{v}_{o''}'(t)$ is known and $\underline{v}_{o'}'(t)$ should be estimated, this may be the case for example for platform-navigation in a stationary environment (here $\underline{v}_{o''}'(t) \equiv \underline{0}$).

In order to keep the presentation compact and with regard to the application treated in chapter 4, we only treat the former case.

The same problem arises for estimation of the rotational motions. Assuming measurements for the relative orientation $\overline{\kappa}(t)$ available, the state-space model (3-22) , (3-17g) together with these measurements is in general not observable, since only the difference of the angular velocities is recoverable from the measurements, see (3-22).

The same remarks as for translational motions apply. In the sequel we assume that the angular velocity of the observer $\underline{\omega}_c(t)$ is known and $\underline{\omega}_b'(t)$ has to be estimated, the differential equations (3-17h) are no longer needed.

For the case considered here, there is in general no knowledge about the rotational accelerations $\underline{b}'_{\omega b}$ in (3-17g) and the transformed higher time derivatives $\underline{\alpha}'_{o''}$ in (3-17f). We assume, however, that the special case described at the end of sec. 3.3 is a reasonable approximation for the real object-movements. Therefore, $\underline{b}'_{\omega b}$ and $\underline{\alpha}'_{o''}$ are sufficiently small and need not be estimated. They are collected into a vector of unknown small disturbances denoted by $\underline{w}$. A vector $\underline{u}$ accounts for the known inputs $\underline{v}_{o'}$ and $\underline{\omega}_c$. Summing up, we consider the state space model of the partial state $\underline{x}$ composed of all quantities to be estimated

$$\dot{\underline{x}} = \underline{f}(\underline{x}, \underline{u}) + \underline{w} \qquad (3\text{-}43)$$

where

$$\underline{x}^T = [\, \underline{r}'^T_{o''}, \underline{v}'^T_{o''}, \underline{b}'^T_{o''}, \overline{\underline{\kappa}}^T, \underline{\omega}'^T_b, \underline{r}''^T_1, \dots, \underline{r}''^T_M \,]$$

$$\underline{u}^T = [\, \underline{v}^T_{o'}, \underline{\omega}^T_c \,]$$

$$\underline{w}^T = [\, \underline{\alpha}'^T_{o''}, \underline{b}'^T_{\omega b} \,]$$

Filter structure

The state vector of the model (3-43) has dimension $n = 16 + 3M$, being rather high especially when the set of structure points M is large. Since the number of required floating-point operations of the Kalman filter algorithms increase with n^3, [MAYBECK 79], it is highly desirable to reformulate the estimation problem in order to decrease the computational burden.

One possibility would be, to devide the whole problem into smaller subproblems and solve each subproblem independently from the others. Since the differential equations for the translational motions, rotational motions and the structure are mutually decoupled, an organization using three independent filter-processes is suggested.

Rotation-filter

The state space model for the rotational motion states, denoted by $\underline{x}_R$, is given by, see (3-22), (3-17g)

$$\underline{\dot{x}}_R = \begin{pmatrix} \bar{\dot{\kappa}} \\ \underline{\dot{\omega}}'_b \end{pmatrix} = \begin{pmatrix} \frac{1}{2}\Gamma(\underline{\omega}'_b - \underline{\omega}_c)\,\bar{\kappa} \\ -\underline{\Omega}\,(\underline{\omega}_c)\,\underline{\omega}_b' \end{pmatrix} + \underline{w}_R =: \underline{f}_R\,(\underline{x}_R,\underline{\omega}_c) + \underline{w}_R \tag{3-44}$$

where $\underline{w}_R$ denotes a stochastic process accounting for the unknown accelerations $\underline{b}_{\omega b}'$ and for the errors in the - assumed to be measured- angular velocities of the observer $\underline{\omega}_c$. We assume that

$$E\,\underline{w}_R(t) \equiv \underline{0} \quad \text{and} \quad E\,\underline{w}_R(t_1)\,\underline{w}_R(t_2) = \underline{Q}_R\,\delta(t_1 - t_2) \tag{3-45}$$

hold, where "E" denotes the expectation operator, $\delta(\cdot)$ stands for Dirac's impulse function and $\underline{Q}_R$ is a nonnegative definite matrix.

It has been shown, in sec. 3.5, how vectors $\underline{\mu}_i$ pointing into the prominent directions $\underline{d}_i'$ can be derived from image features, see eq. (3-40):

$$\underline{\mu}_i = c_i \cdot \underline{d}_i' = sign\,(c_i) \cdot |\,c_i\,| \cdot \underline{d}_i' = sign\,(c_i) \cdot |\,c_i\,| \cdot \tilde{R} \cdot \underline{d}_i'' \tag{3-46}$$

where $|\,c_i\,| = \|\,\underline{\mu}_i\,\|$, $i = 1, ..., N$,

$\underline{d}_i''$ denoting the intrinsic prominent directions defined w.r.t. the object coordinate frame. Adding noise $\underline{n}_{Ri}$ to account for the inaccurate localization of image features, we have at every sampling instant t_k $(k = 0,1, ...)$, $3N$ nonlinear measurement equations for the relative orientation represented by the unit quaternion $\bar{\kappa}$

$$\underline{m}_{Ri}\,(t_k) = \underline{\mu}_i\,(t_k) + \underline{n}_{Ri}\,(t_k) = sign\,(c_i(t_k))\,\|\,\underline{\mu}_i(t_k)\,\|\,\tilde{R}\,(\bar{\kappa}(t_k))\,\underline{d}_i'' + \underline{n}_{Ri}\,(t_k)$$

$$= sign\,(c_i(t_k))\,\|\,\underline{m}_{Ri}\,(t_k)\,\|\,\tilde{R}\,(\bar{\kappa}\,(t_k))\,\underline{d}_i'' + \underline{n}_{Ri}^*\,(t_k) \tag{3-47}$$

where another noise process $\underline{n}_{Ri}^*\,(\,t_k\,)$ has been introduced to account for the substitution of $\|\,\underline{\mu}_i(t_k)\,\|$ by $\|\,\underline{m}_{Ri}\,(t_k)\,\|$ since only $\|\,\underline{m}_{Ri}\,(t_k)\,\|$ is known. As has been explained in sec. 3.5 , the intrinsic prominent directions and the sign of $c_i\,(t_k)$ determining the orientation of the measured vectors $\underline{\mu}_i\,(t_k)$ w.r.t. the directions $\underline{d}_i'(t_k)$, must be known. Thus, we have nonlinear measurements $\underline{m}_{Ri} = \underline{h}\,(\underline{x}_R) + \underline{n}_{Ri}^*$ with known $\underline{h}\,(\,\cdot\,)$, as required by the EKF-formalism.

The EKF processes the incoming flow of measurements to produce an estimate of the state $\hat{\underline{x}}_R\,(t_k/t_k)$ at each sampling instant t_k, being approximately the conditional expectation of $\underline{x}_R\,(t_k)$ based on the measurements $\underline{m}_{Ri}\,(t_j)$, i=1,...N, j=1,...k :

$$\hat{\underline{x}}_R\,(t_k/t_k) \approx E\,(\,\underline{x}_R\,(t_k)\,/\underline{m}_{Ri}\,(t_j)\,;\ i = 1, ... , N\,;\ j = 0, 1, ... , k\,) \tag{3-48}$$

This is done by first computing a prediction of $\underline{x}_R$ (t_k) based on measurements up to time t_{k-1}

$$\hat{\underline{x}}_R (t_k/t_{k-1}) \approx E(\underline{x}_R(t_k)/\underline{m}_{Ri}(t_j)) \; ; \; i=1,\dots,N \; ; \; j=0,1,\dots,k-1) \quad (3\text{-}49)$$

which is obtained by integrating the state equations (3-44) over the sampling interval (t_{k-1}, t_k), starting with the initial value $\hat{\underline{x}}(t_{k-1}/t_{k-1})$

$$\hat{\underline{x}}_R(t_k/t_{k-1}) = \int_{t_{k-1}}^{t_k} \underline{f}_R(\hat{\underline{x}}_R(\tau/t_{k-1}), \underline{\omega}_c(\tau)) \, d\tau \quad (3\text{-}50)$$

The integration is done numerically, the inputs $\underline{\omega}_c$ being supplied by measurements of the observer kinematics according to a higher sampling rate.

In a second step, the predictions are corrected taking into account the new measurements $\underline{m}_{Ri}(t_k)$ to arrive at the a posteriori-estimates (3-48) (filter-update) .

The 3N measurement equations are processed sequentially using an U/D factorization of the covariance matrices to ensure numerical stability, for details see [MAYBECK 79], [MAYBECK 82].

Translation-filter

The state equations for the translational motion states denoted by $\underline{x}_T$ are given by, see (3-17b), (3-17d), (3-17g)

$$\begin{pmatrix} \underline{\dot{r}}'_{o''} \\ \underline{\dot{v}}'_{o''} \\ \underline{\dot{b}}'_{o''} \end{pmatrix} = \begin{pmatrix} -\underline{\Omega}(\underline{\omega}_c)\,\underline{r}'_{o''} + \underline{v}'_{o''} - \underline{v}'_{o'} \\ -\underline{\Omega}(\underline{\omega}_c)\,\underline{v}'_{o''} + \underline{b}'_{o''} \\ -\underline{\Omega}(\underline{\omega}_c)\,\underline{b}'_{o''} \end{pmatrix} + \underline{w}_T \quad (3\text{-}51)$$

where $\underline{w}_T$ denotes a stochastic process accounting for the unknown $\underline{\alpha}'_{o''} = \underline{R}_c\,\bar{\underline{r}}_{o''}$, see (3-17f) and for the errors in the -assumed to be measured- velocity $\underline{v}_{o'}$'. As before it is assumed that

$$E\,\underline{w}_T(t) \equiv \underline{0} \quad \text{and} \quad E\,\underline{w}_T(t_1)\,\underline{w}_T(t_2) = \underline{Q}_T\,\delta(t_1 - t_2)$$

hold, $\underline{Q}_T$ being a nonnegative definite matrix.

The equations (3-26) define the nonlinear relationship between the observer coordinates of an object point P_i and the image coordinates of its projections into the image planes of the stereo system. Considering the object-origin O'' as a special case, there are 4 nonlinear measurement equations for the relative position $\underline{r}'_{o''}$ at each sampling instant t_k .

$$\underline{m}_{o''}(t_k) = \underline{h}_{o''}(\underline{r}'_{o''}(t_k)) + \underline{n}_{To''}(t_k) \tag{3-53}$$

Here, $\underline{n}_{To''}(t_k)$ denotes zero mean, white gaussian noise, that reflects errors in the localization of the projections of O'' .

Since the measurements are corrupted by noise, it is advisable to aggregate more information by incorporating measurements resulting from the M structure points $\underline{r}'_i$, $i = 1, 2, \dots, M$.

By (3-18) we have $\underline{r}'_i = \underline{r}'_{o''} + \tilde{R} \cdot \underline{r}_i''$. Substituting this into (3-26), we get 4M equations relating the 3D-coordinates of O'' to the image coordinates of the projections of the structure points:

$$\frac{x_{li}}{f_l} = \frac{\underline{e}_{1l}^T \underline{r}'_{o''} + \tilde{\underline{e}}_{1l}^T \underline{r}_i'' + T_{xl}}{\underline{e}_{2l}^T \underline{r}'_{o''} + \tilde{\underline{e}}_{2l}^T \underline{r}_i'' + T_{yl}} \qquad \frac{x_{ri}}{f_r} = \frac{\underline{e}_{1r}^T \underline{r}'_{o''} + \tilde{\underline{e}}_{1r}^T \underline{r}_i'' + T_{xr}}{\underline{e}_{2r}^T \underline{r}'_{o''} + \tilde{\underline{e}}_{2r}^T \underline{r}_i'' + T_{yr}}$$

$$\frac{z_{li}}{f_l} = \frac{\underline{e}_{3l}^T \underline{r}'_{o''} + \tilde{\underline{e}}_{3l}^T \underline{r}_i'' + T_{zl}}{\underline{e}_{2l}^T \underline{r}'_{o''} + \tilde{\underline{e}}_{2l}^T \underline{r}_i'' + T_{yl}} \qquad \frac{z_{ri}}{f_r} = \frac{\underline{e}_{3r}^T \underline{r}'_{o''} + \tilde{\underline{e}}_{3r}^T \underline{r}_i'' + T_{zr}}{\underline{e}_{2r}^T \underline{r}'_{o''} + \tilde{\underline{e}}_{2r}^T \underline{r}_i'' + T_{yr}} \tag{3-54}$$

where $\underline{R}_l \tilde{R} =: \begin{pmatrix} \tilde{\underline{e}}_{1l}^T \\ \tilde{\underline{e}}_{2l}^T \\ \tilde{\underline{e}}_{3l}^T \end{pmatrix}$, $\underline{R}_r \tilde{R} =: \begin{pmatrix} \tilde{\underline{e}}_{1r}^T \\ \tilde{\underline{e}}_{2r}^T \\ \tilde{\underline{e}}_{3r}^T \end{pmatrix}$, $i = 1, \dots, M$.

Since the relative rotation $\tilde{R}$ is not known, it may be substituted by its estimate $\hat{\tilde{R}}$ which is a function of $\bar{\kappa}(t_k/t_k)$ (see appendix B) available from the rotation-filter. In the following, we do not explicitly account for errors induced by this substitution, since simulations have shown that estimates for $\tilde{R}$ are in general sufficiently accurate.

(3-54) may be written more compactly

$$\underline{m}_i(t_k) = \underline{h}(\underline{r}'_{o''}(t_k), \underline{r}''_i(t_k)) + \underline{n}_{Ti}(t_k) \; ; \; i = 1, \dots, M \tag{3-55}$$

where $\underline{m}_i(t_k) = \left(\dfrac{x_{li}(t_k)}{f_l}, \dfrac{z_{li}(t_k)}{f_l}, \dfrac{x_{ri}(t_k)}{f_r}, \dfrac{z_{ri}(t_k)}{f_r} \right)^T + \underline{n}_{Ti}(t_k)$

and $\underline{n}_{Ti}(t_k)$ has the same meaning as $\underline{n}_{o''}(t_k)$ in (3-53). Since these measurements depend not only on $\underline{r}'_{o''}(t_k)$ but also on the unknown structure parameters, it would be straightforward to design an EKF for both, translational motions and structure. However, as stated earlier, in order to keep the computational effort reasonable, a separation is suggested. To this end equations (3-55) are linearized around the last estimates of

$r'_{o''}$ and r''_i , namely $\hat{r}'_{o''}(t_{k-1}/t_{k-1})$ and $\hat{r}''_i(t_{k-1}/t_{k-1})$.

Writing $\underline{x}(t_k) := r'_{o''}(t_k)$ and $\underline{x}_i(t_k) := r''_i(t_k)$ and suppressing the time-arguments we have

$$\underline{m}_i = \underline{h}(\underline{x}, \underline{x}_i) + \underline{n}_{Ti} \approx \underline{h}(\hat{\underline{x}}, \hat{\underline{x}}_i) + \underline{C}_i \, \delta\underline{x} + \underline{D}_i \delta\underline{x}_i + \underline{n}_{Ti} \tag{3-56}$$

where $\underline{C}_i := \dfrac{\partial}{\partial \underline{x}} \underline{h} \,|_{\underline{x}=\hat{\underline{x}}, \underline{x}_j=\hat{\underline{x}}_j}$, $\underline{D}_i := \dfrac{\partial}{\partial \underline{x}_i} \underline{h} \,|_{\underline{x}=\hat{\underline{x}}, \underline{x}_j=\hat{\underline{x}}_j}$

$\delta\underline{x} := \underline{x} - \hat{\underline{x}}$, $\delta\underline{x}_i := \underline{x}_i - \hat{\underline{x}}_i$

Now, a separation of translational motion estimation from structure estimation can be achieved by treating the term $\underline{D}_i \, \delta\underline{x}_i$ as additional noise in measurements for the relative position $r'_{o''}$.

Hence, (3-56) becomes

$$\delta\underline{m}_i = \underline{C}_i \, \delta\underline{x} + \tilde{\underline{n}}_i \quad , \quad \tilde{\underline{n}}_i := \underline{n}_{Ti} + \underline{D}_i \, \delta\underline{x}_i \tag{3-57}$$

which together with a linearization of (3-53) is the base of an EKF for the translational motions.

Assuming that $\underline{D}_i \, \delta\underline{x}_i$ is zero mean, white, gaussian noise (this is approximately the case since the $\delta\underline{x}_i$ are the errors in the estimates for the structure parameters obtained from suboptimal filters, discussed below) and uncorrelated with $\underline{n}_{Ti}(t_k)$, the covariance of $\tilde{\underline{n}}_i$ is given by

$$E\,(\tilde{\underline{n}}_i \, \tilde{\underline{n}}_i^T) = \underline{R}_i + \underline{D}_i \, \underline{P}_i \, \underline{D}_i^T \tag{3-58}$$

where $\underline{R}_i = E\,(\tilde{\underline{n}}_i \, \tilde{\underline{n}}_i^T)$, $\underline{P}_i = E\,(\delta\underline{x}_i \, \delta\underline{x}_i^T)$

Similarly, as for the rotation filter, an EKF using U/D factorization is designed to calculate estimates $\hat{\underline{x}}_T(t_k/t_k)$ at every sampling instant. For integrating the state-space equations over each sampling interval (similar as eq. (3-50)), the velocities $\underline{v}'_{o'}$ of the observer are needed.

Estimation of structure

The state equations for the structure parameters $\underline{r}''_i$, $i = 1, \dots, M$ are given by (3-41). Since they are mutually decoupled, an estimator for each structure parameter may be defined independently from the others using equations (3-55) as measurements.

Similarly as in the preceeding paragraph, we consider the linearizations (3-56), but now treating $\underline{C}_i \, \underline{\delta x}_i$ as noise leading to

$$\underline{\delta m}_i = \underline{D}_i \, \underline{\delta x}_i + \underline{n}_i^* \quad ; \quad \underline{n}_i^* := \underline{n}_{Ti} + \underline{C}_i \, \underline{\delta x} \qquad (3\text{-}59)$$

With similar assumptions as above, the covariance of $\underline{n}_i^*$ is given by

$$E \, \underline{n}_i^* \underline{n}_i^{*T} = \underline{R}_i + \underline{C}_i \, \underline{P}_{o^{\prime\prime}} \, \underline{C}_i^T =: \underline{S}_i \qquad (3\text{-}60)$$

where $\underline{P}_{o^{\prime\prime}}$ denotes the covariance of the estimation errors of $\underline{r}'_{o^{\prime\prime}}$, available from the translation filter.

Assuming that the noise $\underline{n}_i^*$ is gaussian, the minimum-variance estimators for the structure parameters $\underline{r}''_i$ are the conditional expectations $\underline{\hat{r}}''_i(t_k/t_k) = E \, (\underline{r}''_i \, (\, t_k \,) \, / \, \underline{m}_i(t_j) \; ; \; j = 1, 2, \dots , k)$, which can be computed recursively

for i=1,..., M :

$$\underline{\hat{r}}''_i(t_k/t_k) = \underline{\hat{r}}''_i(t_{k-1}/t_{k-1}) + \underline{K}_i(t_k) \, [\; \underline{m}_i(t_k) - \underline{D}_i(t_k) \, \underline{\hat{r}}''_i(t_{k-1}/t_{k-1}) \;] \qquad (3\text{-}61a)$$

$$\underline{K}_i(t_k) = \underline{P}_i(t_{k-1}/t_{k-1}) \cdot \underline{D}_i(t_k) \, [\; \underline{S}_i(t_k) + \underline{D}_i(t_k) \, \underline{P}_i(t_{k-1}/t_{k-1}) \, \underline{D}_i^T(t_k) \;]^{-1} (3\text{-}61b)$$

$$\underline{P}_i(t_k/t_k) = \underline{P}_i(t_{k-1}/t_{k-1}) - \underline{K}_i(t_k) \, \underline{D}_i(t_k) \, \underline{P}_i(t_{k-1}/t_{k-1}) \qquad (3\text{-}61c)$$

The recursions may be started with zero initial values and $\underline{P}_i(t_0/t_0)$ diagonal with large elements. However, simulations have shown that instead of using (3-61), a realization relying on an U/D factorization is preferable.

3.7 Results

In order to verify the validity of the presented theory, experiments involving synthetic stereo image sequences have been performed. Fig. 3-5 shows a typical experimental setup.

At initial time $t = t_0$ a stereo setup is placed 0.6m in front of a cube with sidelength 0.1m being the object of interest. The optical axes of the two cameras are parallel, the resolution is 256×256 pixels. The focal lengths are both equal to 7.24mm and the sidelengths of the identical quadratic sensor chips is 6mm. This results in a viewing angle of approximately 45°. The distance of the cameras is 0.1m, the origin of the observer coordinate frame O' lies midway on the line joining the two centers of projection. Hence, the geometrical parameters of the stereo setup are

$$\underline{R}_l = \underline{R}_r = \underline{1} \quad , \quad \underline{T}_l = -\underline{T}_r = \begin{pmatrix} 0.05m \\ 0 \\ 0 \end{pmatrix}$$

The object coordinate frame is defined as depicted in fig. 3-6.

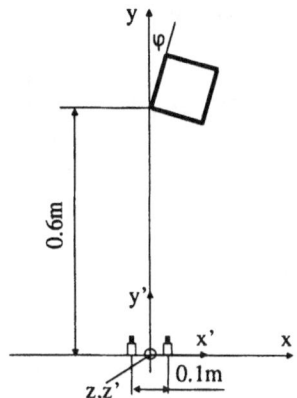

Fig. 3-5 Experimental setup

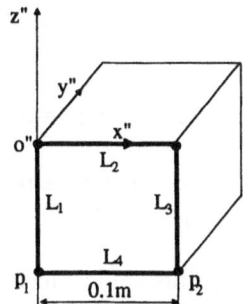

Fig. 3-6 Definition of object

At $t = t_0$ the observer coordinate frame is aligned with the world coordinate frame. The cube is rotated around the z-axis by an angle of $\varphi = -14.9°$. Thus, the relative rotation matrix and the unit quaternion are at $t = t_0$

$$\underline{\tilde{R}}(\underline{\overline{\kappa}}(t_0)) = \begin{pmatrix} 0.9664 & 0.2571 & 0 \\ -0.2571 & 0.9664 & 0 \\ 0 & 0 & 1 \end{pmatrix} \quad ; \quad \underline{\overline{\kappa}}(t_0) = \begin{pmatrix} 0 \\ 0 \\ 0.1296 \\ 0.9916 \end{pmatrix}$$

During the course of the experiment, the cube and the observer perform 3D-motions according to the following velocities (represented w.r.t. the world frame):

object: $\underline{\dot{r}}_{o''}^T \equiv [\,0.4\,,0.4\,,0.2\,]\ m/s \quad ; \quad \underline{\omega}_b^T \equiv [\,0.2\,,0.3\,,0.4\,]\ rad/s$

observer: $\underline{\dot{r}}_{o'}^T \equiv [\,0.4\,,0.45\,,0.2\,]\ m/s \quad ; \quad \underline{\omega}_c^T \equiv [\,0\,,0.2\,,0.1\,]\ rad/s$

The axis of the object-rotation intersects the object-point P_1, see fig. 3-6, and points into the direction of $\underline{\omega}_b$, the axis of the observer-rotation intersects the observer origin and has direction $\underline{\omega}_c$. As can be seen, the observer slowly approaches the object (in the y-direction of the world frame) and otherwise has the same translations as the object, its rotations are completely different from those of the object.

By intersection of the line segments L_1, L_2, L_3, L_4 the three points O", P_1 and P_2 are defined and tracked over time. There are many possible ways to define intrinsic prominent directions, for instance, assuming that it is known a priori that the front face of the object is quadratic, two intrinsic prominent directions are defined

$$\underline{d}"_1 = -\frac{1}{\sqrt{2}} \cdot \underline{e}"_x + \frac{1}{\sqrt{2}} \cdot \underline{e}"_z = \begin{pmatrix} -\frac{1}{\sqrt{2}} \\ 0 \\ \frac{1}{\sqrt{2}} \end{pmatrix} \quad , \quad \underline{d}"_2 = \underline{e}"_z = \begin{pmatrix} 0 \\ 0 \\ 1 \end{pmatrix},$$

$\underline{d}"_1$ being the direction joining the points P_2, O".

The calculation of the 2D-straight line parameters according to (3-36) is based on the images of the points O", P_1, P_2.

The image points are localized only with pixel-accuracy giving rise to noise which is characteristic for stereoscopic measurements.

Note, that the relative depth of the object (y'- component of the relative position $\underline{r}'_{o"}$) varies between 0.4m and 0.6m which is four to six times the stereo base.

The exact values of the observer velocities $\underline{\dot{r}}_{o'}$ and $\underline{\omega}_c$ were provided to the EKF's. The sampling time was selected to be T = 0.1s, so that every 0.1s a stereo image has been evaluated.

Fig. 3-7 - 3-21 show the results of the experiment vs. time in seconds. In fig. 3-7 - 3-19 the full lines denote true quantities, dotted lines denote estimated quantities. In fig. 3-20, 3-21 the true values are $\underline{r}_1" = [0, 0, -0.1]^T$ and $\underline{r}_2" = [0.1, 0, -0.1]^T$ respectively, see fig. 3-6. As can be seen, the various estimated quantities converge to their respective true values.

Note, that the y'-components of $\underline{r}'_{o"}$ and $\underline{v}'_{o"}$ are estimated with lower accuracy than the other components. This is due to an inherent feature of stereoscopic measurements, reflecting the fact that the errors in the measurements of depth are larger than those in the x' and z' coordinates. For the same reason, the 1. and 3. component of the estimated quaternion $\overline{\kappa}$ (representing the components of relative rotations about the x' and z' axes) and the x' and z' components of the estimated angular velocity $\underline{\omega}'_b$ are less accurate than the y' components.

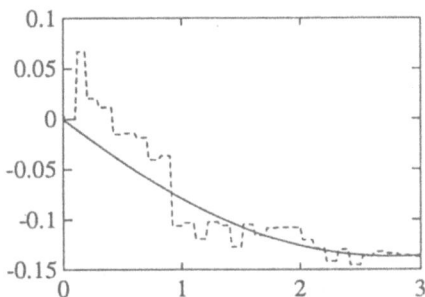

Fig. 3-7 1. component of unit quaternion

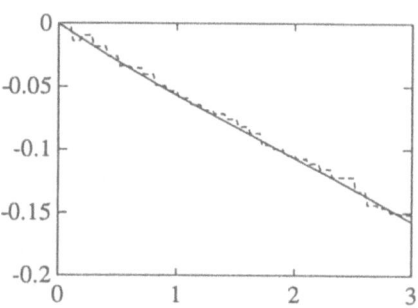

Fig. 3-8 2. component of unit quaternion

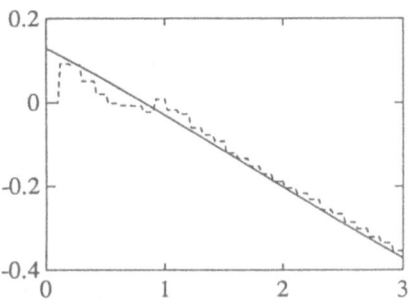

Fig. 3-9 3. component of unit quaternion

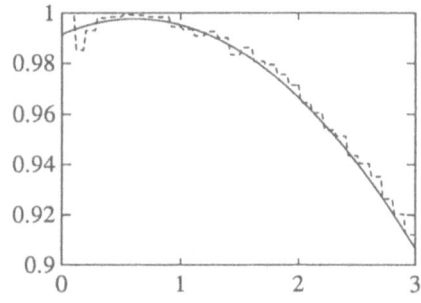

Fig.3-10 4. component of unit quaternion

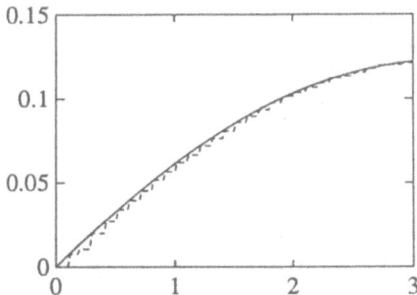

Fig. 3-11 x'-component of relative position $\underline{r}'_{o}$"

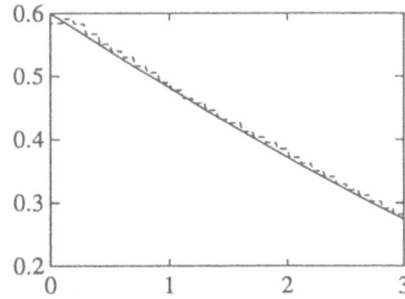

Fig. 3-12 y'-component of relative position $\underline{r}'_{o}$"

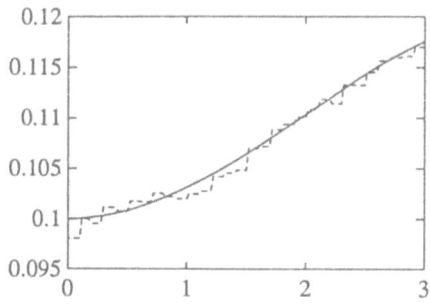

Fig. 3-13 z'-component of relative position
          $\underline{r}'_o$"

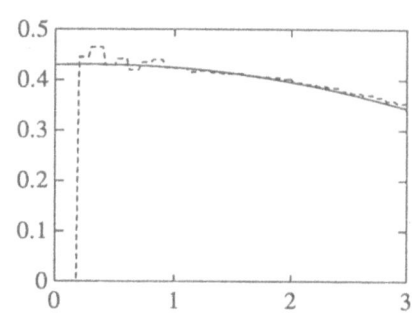

Fig. 3-14 x'-component of linear velocity of
          object $\underline{v}'_o$"

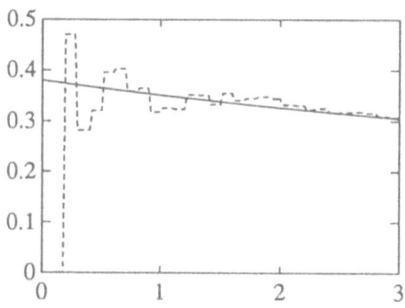

Fig. 3-15 y'-component of linear velocity of
          object $\underline{v}'_o$"

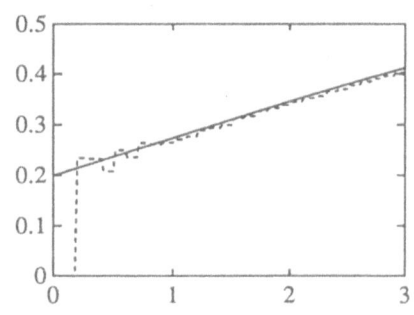

Fig. 3-16 z'-component of linear velocity of
          object $\underline{v}'_o$"

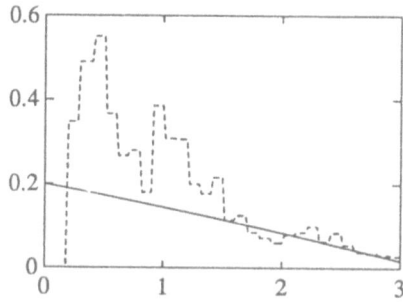

Fig. 3-17 x'-component of angular velocity
          of object $\underline{\omega}'_b$

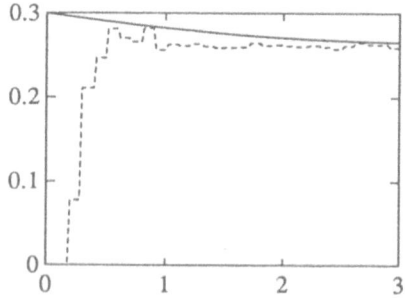

Fig. 3-18 y'-component of angular velocity of
          object $\underline{\omega}'_b$

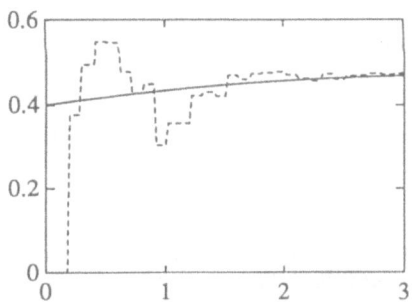

Fig. 3-19 z'-component   of   angular velocity of object $\underline{\omega}'_b$

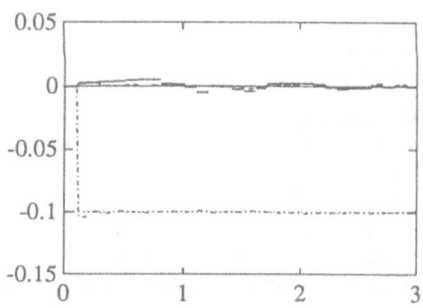

Fig. 3-20 Coordinates of structure point P_1

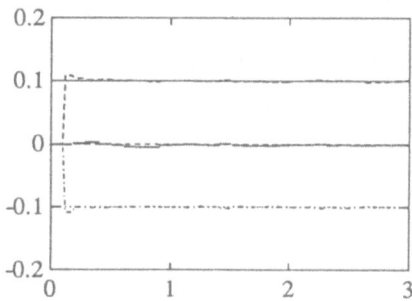

Fig. 3-21 Coordinates of structure point P_2

3.8 References

[AYACHE, FAVERJON 87]
Ayache, N.; Faverjon, B.: "Efficient Registration of Stereo Images by Matching Graph Descriptions of Edge Segments
The International Journal of Computer Vision, 1(2), April, 1987

[BOECKER et al. 86]
Boecker ; Hartmann, I.; Zwanzig, Ch.: "Nichtlineare und Adaptive Regelungssysteme"
Springer Verlag, Berlin, 1986

[COURANT, HILBERT 63]
Courant, R.; Hilbert, D.: "Methods of Mathematical Physics"
Interscience Publishers, New York, Vol.1, Forth Printing , 1963, pp. 536-539

[DUDA, HART 73]
Duda, R.O.; Hart, P.E.: Pattern Classification and Scene Analysis"
John Wiley and Sons, New York, 1973

[GRIMSON 81]
Grimson, W.E.L.: "From Images to Surfaces: A Computational Study of the Human Early Visual System"
MIT Press, Cambridge, Mass., 1981

[HARALICK 84]
Haralick, R.M.: "Digital Step Edges from Zero Crossing of Second Order Directional Derivatives"
IEEE Trans. on Pattern Analysis and Machine Intelligence, PAMI-6, 1, 1984, pp. 58-68

[HOBROUGH, HOBROUGH 83]
Hobrough, G.; Hobrough, T.: "Stereopsis for Robots by Iterative Stereo Matching"
Third International Conference on Robot Vision and Sensory Controls, SPIE Cambridge Symposium on Optical and Electro-Optical Engineering, 1983, pp. 94-102

[HORN 87]
Horn, B.K.P.: "Robot Vision"
The MIT Elecrical Engineering and Computer Science Series, MIT Press, 1987

[HUANG, BLOSTEIN 85]
Huang, T.S.; Blostein, S.D.: "Robust Algorithms for Motion Estimation Based on Two Sequential Stereo Image Pairs"
Proc. International Conference on Pattern Recognition, 1985, pp. 518-523

[MARR, POGGIO 79]
Marr, D.; Poggio, T.: "A Computational Study of Human Stereo Vision"
Proc. Royal Society London B. 204, 1979, pp. 301-328

[MAYBECK 79]
Maybeck, P.S.: "Stochastic Models, Estimation and Control"
Vol. 1, Academic Press, New York, 1979

[MAYBECK 82]
Maybeck, P.S.: "Stochastic Models, Estimation and Control"
Vol. 2, Academic Press, New York, 1982

[MORAVEC 79]
Moravec, H.P.: "Visual Mapping by a Robot Rover"
Proc. Sixth International Conference on Artificial Intelligence, Tokyo, 1979, pp. 598-600

[NAGEL, NEUMANN 81]
Nagel, H.H.; Neumann, B.: "On 3D-Reconstruction from Two Perspective Views"
Proc. International Joint Conference on Artificial Intelligence, Vancouver/BC, Canada, 1981, pp. 661-663

[NISHIHARA 83]
Nishihara, H.K.: "PRISM: A Practical Real Time Imaging Stereo Matcher"
Third International Conference on Robot Vision and Sensory Controls, SPIE Cambridge Symposium on Optical and Electro-Optical Engineering, 1983

[PAUL 81]
Paul, R.P.: "Robot Manipulator: Mathematics, Programming and Control"
MIT Press, Cambridge, Mass., 1981

[POLLARD et al. 85]
Pollard, S.B.; Mayhew, J.E.W.; Frisby, J.P.: "PMF: A Stereo Correspondence Algorithm Using a Disparity Gradient Limit"
Perception, 14, 1985, pp. 449-470

[PUSCH 90]
Pusch, M.: "3D-Erfassung der Umwelt einer mobilen Plattform aus Stereobildern"
Diplomarbeit am Institut fuer Regelungstechnik und Systemdynamik der TU-Berlin, 1990

[ROACH, AGGARWAL 80]
Roach, J.W.; Aggarwal, J.K.: "Determining the Movements of Objects from a Sequence of Images"
IEEE Trans. on Pattern Analysis and Machine Intelligence, PAMI-2, 6, 1980, pp. 554-562

[SPRING 86]
Spring, K.W.: "Euler Parameters and the Use of Quaternion Algebra in the Manipulation of Finite Rotations: a Review"
Mechanism and Machine Theory, Vol. 21, No. 5, 1986, pp. 365-373

[STUELPNAGEL 64]
Stuelpnagel, J.: "On the Parametrization of the Three-Dimensional Rotation Group"
SIAM Review, 6, 1964, pp.422-430

[TSAI, HUANG 84]
Tsai, R.Y.; Huang, T.S.: "Uniqueness and Estimation of Three-Dimensional Motion Parameters of Rigid Objects with Curved Surface"
IEEE Trans. on Pattern Analysis and Machine Intelligence, PAMI-6, 1, 1984, pp. 13-26

[VAN TREES 68]
Van Trees, H.L.: "Detection, Estimation and Modulation Theory, Part I"
John Wiley and Sons, New York, 1968

[WALL, DANIELSSON 84]
Wall, K.; Danielsson, P.E.: "A Fast Sequential Method for Polygonal Approximation of Digitized Curves"
Computer Vision, Graphics and Image Processing, 28, 1984, pp. 220-227

[WU et al. 88]
Wu, J.J.; Rink, R.E.; Caelli, T.M.; Gourishankar, V.G.: "Recovery of the 3D-Location and Motion of a Rigid Object Through Camera Image (An Extended Kalman Filter Approach)"
International Journal of Computer Vision, 3, (1988), pp. 373-394

[YAKIMOVSKY, CUNNINGHAM 78]
Yakimovsky, Y.; Cunningham, R.: "A System for Extracting Three-Dimensional Measurements from a Stereo Pair of TV Cameras"
Computer Graphics and Image Processing, 7, 1978, pp. 195-210

[YOUNG, CHELLAPPA 88]
Young, G.S.; Chellappa, R.: "3-D Motion Estimation Using a Sequence of Noisy Stereo Images"
Proc. IEEE Conference on Computer Vision and Pattern Recognition, 1988, pp. 710-716

Appendix A : Proof of elementary theorems

Proposition 1

Consider a coordinate frame moving relative w.r.t. the world coordinate frame, the coordinate transformation of an arbitrary point P with position vector $\underline{r}_p$ being

$$\underline{r}'_p = \underline{R}_c\,(\,\underline{r}_p - \underline{r}_{o'}\,) \tag{A1}$$

where $\underline{r}'_p$ denotes the position vector of P w.r.t. the moving coordinate frame, $\underline{R}_c$ being a rotation matrix and $\underline{r}_{o'}$ denoting the position vector of the origin.

Then the time-derivative of the transformed vector $\underline{\dot{r}}'_p$ is given by

$$\underline{\dot{r}}'_p = -\,\underline{\omega}_c \times \underline{r}'_p + \underline{R}_c\,(\,\underline{\dot{r}}_p - \underline{\dot{r}}_{o'}\,) \tag{A2}$$

where $\underline{\omega}_c$ denotes the angular velocity of the moving coordinate frame according to (3-2).

Proof

Taking the time-derivative of (A1) gives

$$\underline{\dot{r}}'_p = \underline{\dot{R}}_c\,(\,\underline{r}_p - \underline{r}_{o'}\,) + \underline{R}_c\,(\,\underline{\dot{r}}_p - \underline{\dot{r}}_{o'}\,) = \underline{\dot{R}}_c\,\underline{R}_c^T\,\underline{r}'_p + \underline{R}_c\,(\,\underline{\dot{r}}_p - \underline{\dot{r}}_{o'}\,) \tag{A3}$$

Now, consider a point P' rigidly connected with the moving coordinate system. (A1) gives

$$\underline{r}_{p'} = \underline{R}_c^T\,\underline{r}'_{p'} + \underline{r}_{o'}$$

$$\Rightarrow \quad \underline{\dot{r}}_{p'} = \underline{\dot{R}}_c^T\,\underline{r}'_{p'} + \underline{R}_c\,\underline{\dot{r}}'_{p'} + \underline{\dot{r}}_{o'}$$

Here, $\underline{\dot{r}}'_{p'} = \underline{0}$ because P' is rigidly connected with the moving coordinate frame, thus, it follows

$$\underline{\dot{r}}_{p'} = \underline{\dot{R}}_c^T\,\underline{r}'_{p'} + \underline{\dot{r}}_{o'} \tag{A4}$$

On the other hand, the velocity of P' is given by

$$\dot{r}_{p'} = \dot{r}_{o'} + \underline{\omega}_c \times (r_{p'} - r_{o'})$$

Comparing with (A4) gives

$$\underline{\dot{R}}_c^T r'_{p'} = \underline{\omega}_c \times (r_{p'} - r_{o'})$$

or with (A1)

$$\underline{\dot{R}}_c^T \underline{R}_c (r_{p'} - r_{o'}) = \underline{\omega}_c \times (r_{p'} - r_{o'})$$

Since P' is arbitrary, we have the relation

$$\underline{\dot{R}}_c^T \underline{R}_c \, r = \underline{\omega}_c \times r$$

Now, if the moving coordinate system rotates in the inverse sense, $\underline{\omega}_c$ must be replaced by $-\underline{\omega}_c$ and $\underline{R}_c$ by $\underline{R}_c^{-1} = \underline{R}_c^T$.

This gives the relation

$$\underline{\dot{R}}_c \underline{R}_c^T \, r = -\underline{\omega}_c \times r \tag{A5}$$

Applying this result to (A3) gives (A2).   ∎

Proposition 2

Given the canonical orthonormal basis ($\underline{i} = [1,0,0]^T$; $\underline{j} = [0,1,0]^T$; $\underline{k} = [0,0,1]^T$) defining a right-hand reference coordinate frame and a rotated version ($\underline{e}_1$, $\underline{e}_2$, $\underline{e}_3$). Let $\underline{R}$ denote the rotation matrix defining the relationship between the two coordinate frames, i.e.

$$\underline{i} = \underline{R}\,\underline{e}_1 \quad , \quad \underline{j} = \underline{R}\,\underline{e}_2 \quad , \quad \underline{k} = \underline{R}\,\underline{e}_3$$

$$\Rightarrow \quad \underline{R}^T = [\,\underline{e}_1, \underline{e}_2, \underline{e}_3\,] \quad ; \quad \underline{R} = \begin{pmatrix} \underline{e}_1^T \\ \underline{e}_2^T \\ \underline{e}_3^T \end{pmatrix} \tag{A-6}$$

Then, given two arbitrary vectors $\underline{a}$, $\underline{b}$,

$$\underline{R}\,(\underline{a} \times \underline{b}) = \underline{R}\,\underline{a} \times \underline{R}\,\underline{b} \tag{A-7}$$

holds.

Proof

The left-hand side of (A-6) can be rewritten

$$\underline{R}\,(\underline{a}\times \underline{b}) = \underline{R}\,\underline{\Omega}(\underline{a})\,\underline{b} = \underline{R}\,\underline{\Omega}(\underline{a})\,\underline{R}^T\,\underline{R}\,\underline{b} \qquad \text{(A-8)}$$

where

$$\underline{\Omega}(\underline{a}) = \begin{pmatrix} 0 & -a_z & a_y \\ a_z & 0 & -a_x \\ -a_y & a_x & 0 \end{pmatrix}$$

Using (A-6), we have

$$\underline{\Omega}(\underline{a})\,\underline{R}^T = [\,\underline{a}\times \underline{e}_1\,,\underline{a}\times \underline{e}_2\,,\underline{a}\times \underline{e}_3\,]$$

$$\Rightarrow \quad \underline{R}\,\underline{\Omega}(\underline{a})\,\underline{R}^T = \begin{pmatrix} \underline{e}_1^T(\underline{a}\times \underline{e}_1) & \underline{e}_1^T(\underline{a}\times \underline{e}_2) & \underline{e}_1^T(\underline{a}\times \underline{e}_3) \\ \underline{e}_2^T(\underline{a}\times \underline{e}_1) & \underline{e}_2^T(\underline{a}\times \underline{e}_2) & \underline{e}_2^T(\underline{a}\times \underline{e}_3) \\ \underline{e}_3^T(\underline{a}\times \underline{e}_1) & \underline{e}_3^T(\underline{a}\times \underline{e}_2) & \underline{e}_3^T(\underline{a}\times \underline{e}_3) \end{pmatrix} \quad \text{(A-9)}$$

Since for arbitrary vectors $\underline{x}\,,\underline{y}\,,\underline{z}$, the scalar triple product $\underline{x}^T(\underline{y}\times \underline{z})$ satisfies

$$\underline{x}^T(\underline{y}\times \underline{z}) = \underline{y}^T(\underline{z}\times \underline{x}) = -\underline{y}^T(\underline{x}\times \underline{z}) \quad ,$$

and for a right-hand system of unit vectors we have

$$\underline{e}_1\times \underline{e}_2 = \underline{e}_3 \quad , \quad \underline{e}_2\times \underline{e}_3 = \underline{e}_1 \quad , \quad \underline{e}_3\times \underline{e}_1 = \underline{e}_2 \quad ,$$

(A-9) results in

$$\underline{R}\,\underline{\Omega}(\underline{a})\,\underline{R}^T = \begin{pmatrix} 0 & -\underline{a}^T\underline{e}_3 & \underline{a}^T\underline{e}_2 \\ \underline{a}^T\underline{e}_3 & 0 & -\underline{a}^T\underline{e}_1 \\ -\underline{a}^T\underline{e}_2 & \underline{a}^T\underline{e}_1 & 0 \end{pmatrix} = \underline{\Omega}\begin{pmatrix} \underline{a}^T\underline{e}_1 \\ \underline{a}^T\underline{e}_2 \\ \underline{a}^T\underline{e}_3 \end{pmatrix} = \underline{\Omega}\begin{pmatrix} \underline{e}_1^T\underline{a} \\ \underline{e}_2^T\underline{a} \\ \underline{e}_3^T\underline{a} \end{pmatrix} \quad \text{(A-10)}$$

Substitution of (A-10) into (A-8) and using (A-6) gives (A-7) :

$$\underline{R}(\underline{a}\times \underline{b}) = \underline{\Omega}\begin{pmatrix} \underline{e}_1^T\underline{a} \\ \underline{e}_2^T\underline{a} \\ \underline{e}_3^T\underline{a} \end{pmatrix}\cdot \underline{R}\,\underline{b} = \underline{R}\,\underline{a}\times \underline{R}\,\underline{b} \qquad \blacksquare$$

Remark

In a similar way, (A-7) can also be proved in case $\underline{R}$ is represented w.r.t. a left-hand coordinate frame.

Appendix B : Representation of rotations by unit quaternions

Let $\underline{R}$ denote a rotation matrix, i.e. $\underline{R}\,\underline{R}^T = \underline{I}$ ($\underline{I}$ denoting the identity matrix). Given any real numbers κ_i ($i = 1, 2, 3, 4$) for which $(\kappa_1^2 + \kappa_2^2 + \kappa_3^2 + \kappa_4^2)^{1/2} = v > 0$ and let $w := (\kappa_1^2 + \kappa_2^2 + \kappa_3^2)^{1/2}$ then Caley's Theorem, see [COURANT, HILBERT 63] states that

-The matrix $\underline{R} = \dfrac{1}{v^2}\begin{pmatrix} \kappa_4^2 + \kappa_1^2 - \kappa_2^2 - \kappa_3^2 & 2(\kappa_1\kappa_2 - \kappa_3\kappa_4) & 2(\kappa_1\kappa_3 + \kappa_2\kappa_4) \\ 2(\kappa_1\kappa_2 + \kappa_3\kappa_4) & \kappa_4^2 + \kappa_2^2 - \kappa_1^2 - \kappa_3^2 & 2(\kappa_2\kappa_3 - \kappa_1\kappa_4) \\ 2(\kappa_1\kappa_3 - \kappa_2\kappa_4) & 2(\kappa_2\kappa_3 + \kappa_1\kappa_4) & \kappa_4^2 + \kappa_3^2 - \kappa_1^2 - \kappa_2^2 \end{pmatrix}$ (B-1)

is orthogonal, the vector $\underline{n}^T = [\, \kappa_1/w, \kappa_2/w, \kappa_3/w \,]$ is a unit vector pointing into the direction of the rotation axis,

-if θ denotes the angle of rotation, then

$$\cos\frac{\theta}{2} = \frac{\kappa_4}{v} \quad , \quad \sin\frac{\theta}{2} = \frac{w}{v}$$ (B-2)

-in case that $v = 1$, then any two sets κ_i, $\tilde{\kappa}_i$ define the same $\underline{R}$ if and only if $\kappa_i = \tilde{\kappa}_i$ or $\kappa_i = -\tilde{\kappa}_i$ $i = 1, 2, 3, 4$.

The proof may be found in [COURANT, HILBERT 63].

Now, consider 4- tuples called "quaternions"

$$\underline{\kappa} = \begin{pmatrix} \underline{\kappa}_v \\ \kappa_s \end{pmatrix} \qquad \underline{\kappa}_v \in R^3 \ , \ \kappa_s \in R$$ (B-3)

where $\underline{\kappa}_v$ denotes the "vector part" of a quaternion and κ_s its "scalar part". Quaternions may also be viewed as "supercomplex numbers" having a real part (κ_s) and three imaginary parts ($\kappa_{v1}, \kappa_{v2}, \kappa_{v3}$), see [HORN 87], but for our purposes the representation by 4-vectors is more suitable.

For quaternions an algebra is defined through the following properties, [SPRING 86]

-addition: $\underline{\overline{\kappa}}_1 + \underline{\overline{\kappa}}_2 = \begin{pmatrix} \underline{\kappa}_{v1} + \underline{\kappa}_{v2} \\ \kappa_{s1} + \kappa_{s2} \end{pmatrix}$

-multiplication: $\underline{\overline{\kappa}}_1 \otimes \underline{\overline{\kappa}}_2 = \begin{pmatrix} \kappa_{1s}\underline{I} + \underline{\Omega}\,(\underline{\kappa}_{1v}) & \underline{\kappa}_{1v} \\ -\underline{\kappa}_{1v}^T & \kappa_{1s} \end{pmatrix} \cdot \begin{pmatrix} \underline{\kappa}_{2v} \\ \kappa_{2s} \end{pmatrix}$

where "." denotes the usual matrix multiplication

and $\underline{\Omega}(\underline{a}) \cdot \underline{b} := \underline{a} \times \underline{b}$

-identity-quaternion: $\underline{\overline{I}}$: $\underline{1}_v = \underline{0}$, $1_s = 1$

-conjugate quaternion: $\underline{\overline{\kappa}}^+ = \begin{pmatrix} -\underline{\kappa}_v \\ \kappa_s \end{pmatrix}$

-norm: $||\underline{\overline{\kappa}}|| = \underline{\kappa}_v^T \cdot \underline{\kappa}_v + \kappa_s^2$

-inverse: $\underline{\overline{\kappa}}^{-1}$: $\underline{\overline{\kappa}}^{-1} \otimes \underline{\overline{\kappa}} = \underline{\overline{I}}$, $\underline{\overline{\kappa}}^{-1} = \dfrac{1}{||\underline{\overline{\kappa}}||} \cdot \underline{\overline{\kappa}}^+$

In case that $||\underline{\overline{\kappa}}|| = 1$, $\underline{\overline{\kappa}}$ is called a "unit quaternion".

Consider special unit quaternions of the form

$$\underline{\overline{\kappa}} = \begin{pmatrix} \sin\dfrac{\theta}{2}\,\underline{n} \\ \cos\dfrac{\theta}{2} \end{pmatrix} , \quad \underline{n} \in R^3 , \quad ||\underline{n}|| = 1 , \quad \theta \in R \tag{B-4}$$

Then by Caley's Theorem we see that these unit quaternions may represent rotation matrices and therefore arbitrary rotations, where the unit vector $\underline{n}$ points into the direction of the rotation axis and θ denotes the rotation angle. To be more precise, let $\underline{r}^A$ denote a vector expressed w.r.t. a reference frame A and let $\underline{r}^B$ denote the same vector, but expressed w.r.t. a rotated frame B, then we have

$$\underline{r}^A = \underline{R}\,\underline{r}^B \tag{B-5}$$

and $\underline{\overline{\kappa}}^T = [sin(\dfrac{\theta}{2})\,\underline{n}^T , cos(\dfrac{\theta}{2})]$ represents $\underline{R}$ according to (B-1) , when θ and $\underline{n}$ denote the angle and axis of rotation respectively.

Moreover, one can show that the relation (B-5) can be expressed using quaternion algebra

$$\underline{\overline{r}}^A = \underline{\overline{\kappa}} \otimes \underline{\overline{r}}^B \otimes \underline{\overline{\kappa}}^* \ , \quad where \ \underline{\overline{r}}^A = \begin{pmatrix} r^A \\ 0 \end{pmatrix}, \ \underline{\overline{r}}^B = \begin{pmatrix} r^B \\ 0 \end{pmatrix} \tag{B-6}$$

In order to describe rigid body motion, the relationship between the angular velocity of the rigid body and the time derivative of a unit quaternion representing the orientation of the rigid body w.r.t. a reference frame is of interest.

Following an elegant derivation due to [SPRING 86] , consider a rotating rigid body with instantaneous angular velocity $\underline{\omega}$. Let $\underline{R}(t)$ denote the rotation matrix expressing the relationship between the body coordinate frame B and a reference frame A and $\overline{\underline{\kappa}}(t)$ a unit quaternion belonging to $\underline{R}(t)$, according to $\underline{r}^A = \underline{R}\,\underline{r}^B$.

The orientation of the rigid body at time $t + dt$ is represented by $\underline{R}(t + dt) = \underline{R}(t) \cdot \underline{R}(dt)$ and $\overline{\underline{\kappa}}(t + dt) = \overline{\underline{\kappa}}(t) \otimes \overline{\underline{\kappa}}(dt)$ where $\overline{\underline{\kappa}}(dt)$ represents $\underline{R}(dt)$. Hence, for an infinitesimal dt, we have

$$\overline{\underline{\kappa}}(dt) = \begin{pmatrix} \sin\dfrac{\|\underline{\omega}\|\,dt}{2} \cdot \dfrac{\underline{\omega}\,dt}{\|\underline{\omega}dt\|} \\ \cos\dfrac{\|\underline{\omega}\|\,dt}{2} \end{pmatrix} = \begin{pmatrix} \dfrac{1}{2}\underline{\omega}\,dt \\ 1 \end{pmatrix} \tag{B-7}$$

The increment of $\overline{\underline{\kappa}}$ is

$$d\overline{\underline{\kappa}} = \overline{\underline{\kappa}}(t + dt) - \overline{\underline{\kappa}}(t) = \overline{\underline{\kappa}}(t) \otimes \overline{\underline{\kappa}}(dt) - \overline{\underline{\kappa}}(t)$$

$$= \overline{\underline{\kappa}}(t) \otimes [\,\overline{\underline{\kappa}}(dt) - \underline{I}\,] = \overline{\underline{\kappa}}(t) \otimes \begin{pmatrix} \dfrac{1}{2}\underline{\omega}\,dt \\ 0 \end{pmatrix} \tag{B-8}$$

This results in

$$\dot{\overline{\underline{\kappa}}}(t) = \frac{1}{2}\overline{\underline{\kappa}}(t) \otimes \begin{pmatrix} \underline{\omega} \\ 0 \end{pmatrix} \tag{B-9}$$

Carrying out the quaternion multiplication yields

$$\dot{\overline{\underline{\kappa}}}(t) = \begin{pmatrix} \dot{\underline{\kappa}}_v(t) \\ \dot{\kappa}_s(t) \end{pmatrix} = \frac{1}{2}\underline{\Gamma}(\underline{\omega}) \cdot \begin{pmatrix} \underline{\kappa}_v \\ \kappa_s \end{pmatrix} \tag{B-10}$$

where $\underline{\Gamma}(\omega) = \begin{pmatrix} -\underline{\Omega}(\omega) & \underline{\omega} \\ -\underline{\omega}^T & 0 \end{pmatrix}$

Appendix C Algorithm for stereo matching by 3D-line segment chaining

This version of the stereo matcher has been implemented in Parallel C.

Inputs:

- two lists of 2D-line segments, one for each image of the stereo system,

- maximum depth tolerance ε

- allowed range of disparities according to the definition of the working range of the stereo setup v_{min} , v_{max}

1. step: Definition of potential homologous segments

- for each segment S_{li} in the left image:

    - find all potential homologous segments in the right image according to the following constraints:

        a. epipolar constraint
        b. rough coincidence of orientations
        c. rough coincidence of lengths
        d. check whether disparities of endpoints are within allowed range [v_{min} , v_{max}]

result: For each S_{li} : list of potential homologous right-segments

$$i \rightarrow I_i = \{ j \, / \, S_{li} \rightarrow S_{rj} \}$$

- I_i may be empty

- all potential disparities are available.

2. step: Generation of start-hypothesis

- for each segment S_{li} :

  - from the list of potential homologous right segments take the one that maximizes a certain "fitting score" M (i,j):

$$S_{rj} : j \in I_i \; ; \; M(i,j) \geq M(i,k) \text{ for } all \; k \in I_i \; , \; k \neq j$$

result: List of best fitting pairs (S_{li} , S_{rj}), each constituting a "start-hypothesis".

3. step: Chaining of 3D-segments

- for each best fitting pair (S_{li} , S_{rj})

  - (S_{li} , S_{rj}) constitutes a "match": $M(i,j) = (S_{li} , S_{rj})$

  Beginning with the start-hypothesis (S_{li} , S_{rj}) a chain of connected 3D-line segments is constructed, each 3D-line segment being defined by a left-right match of 2D-line-segments. The chain is named Ch_i and initialized with the first match: $Ch_i \leftarrow \{M (i, j)\}$. In order to connect more matches to the chain, the recursive procedure "chain" is called:

  - call procedure chain ((S_{li} , S_{rj}))

    procedure chain (($S_{l\lambda}$, $S_{r\mu}$))

    - $S_{l\lambda}$ defines two neighborhoods $N_1 (\lambda)$, $N_2 (\lambda)$ around the endpoints of $S_{l\lambda}$

    - associated with the pair ($S_{l\lambda}$, $S_{r\mu}$) are

      - the disparities of the pair ($S_{l\lambda}$, $S_{r\mu}$) : $v_1 (\lambda,\mu)$, $v_2(\lambda,\mu)$

      - tolerances of disparities: $| dv_1(\varepsilon ,\lambda,\mu)| , | dv_2(\varepsilon ,\lambda,\mu)|$

    - for all S_{ln} which have at least one endpoint in $N_1 (\lambda)$:

      - check all S_{rm} , $m \in I_n$, whether the resulting disparity $v_1(n,m)$ satisfies the continuity constraint:

$$v_1 (\lambda,\mu) - | dv_1(\varepsilon ,\lambda,\mu)| \leq v_1 (n,m) \leq v_1 (\lambda,\mu) + | dv_1(\varepsilon ,\lambda,\mu)|$$

- if so, then the pair (S_{ln} , S_{rm}) constitutes a match $M(n,m)$ which is added to the chain

$$Ch_\lambda \leftarrow Ch_\lambda \cup \{M(n, m)\}$$

- call procedure chain ((S_{ln} , S_{rm}))

- for all S_{ln} which have at least one endpoint in N_2 (λ) :

  - check all S_{rm} , $m \in I_n$, whether the resulting disparity $v_2(n,m)$ satisfies the continuity constraint:

  $$v_2(\lambda,\mu) - |dv_2(\varepsilon,\lambda,\mu)| \leq v_2(n,m) \leq v_2(\lambda,\mu) + |dv_2(\varepsilon,\lambda,\mu)|$$

  - if so, then the pair (S_{ln} , S_{rm}) constitutes a match $M(n,m)$ which is added to the chain

  $$Ch_\lambda \leftarrow Ch_\lambda \cup \{M(n, m)\}$$

  - call procedure chain ((S_{ln} , S_{rm}))

  - else return

result: List of chains Ch_i, one for each segment S_{li} of the left image:

- each chain contains matches of line segments which are connected in 3D-space according to the given maximum depth-tolerance ε at its connections

- the number of matches contained in a chain is called the "length" of the chain

- if no chaining was possible for an S_{li} , then its chain has length zero.

4. step: Verification of matches

- rename the indices of the chains according to their lengths:

  $$Ch_1 \geq Ch_2 \geq \ldots$$

- set i=1, A = empty, for all i:

  - for all matches $M(n, m)$ contained in Ch_i :

- if S_{ln} is not contained in A then add S_{ln} to A : $A \leftarrow A \cup \{n\}$

- else delete Ch_i

result: List of non overlapping chains, i.e. each segment S_{ln} or S_{rm} that has been matched appears only once in the list of chains (uniqueness-constraint).

4 Dynamic 3D Vision : The Visually Controlled Robot

A. Graffunder, Z. Ren

4.1 Introduction

In this chapter we are interested in how the concepts of chapter 3 may be utilized for the purpose of robot control in closed visual loop. Specifically, we are interested in formulating a control strategy that is suitable for controlling the movements of a robot relative to relevant objects in its environment. This is obvious, since achieving some degree of "autonomy" for a mobile agent (robot) includes achieving some capabilities for situation-dependent movement-behaviour. In many cases, this can be done best by exploiting information from visual sensors.

Mobile platform navigation is perhaps the most prominent example, where visual sensing of the enviroment has been recognized of fundamental importance, thinking of capabilities such as "obstacle avoidance", see e.g. [LOZANO-PEREZ 90], [REM-BOLD 88] and also chapters 6 and 7 of this book.

However, in the field of manipulator control too, there is a need for visual feedback control.

Although the ideas presented in this chapter also apply to mobile platform navigation, we will concentrate on visual feedback control of robotic manipulators.

Applications relate f. i. to the automation of mechanical assembly. According to [WHITNEY 89] a typical operating cycle consists of, (see also [REMBOLD 88])

> gross motion to a part,
> fine motion to fetch it,
> gross motion to bring the part to the assembly point,
> fine motion to insert or place it.

Here, "gross motion" is characterized by only little demands on accurately following a certain (prespecified) path in the workspace of the manipulator, high speed is desired and obstacles have to be avoided. On the other hand, "fine motion" takes place near the part to be manipulated and in contact with the part. The emphasis is on accuracy (thinking of the insertion of a peg into a hole with a small chamfer), although high speeds are also desired, in order to keep the cycle time small.

Nowadays the use of robotic manipulators is dominated by the classical "hard auto-mation" robot designs which are only able to perform preprogramed actions on assembly lines, since no external sensing of their environment is envolved. Thus, all details of an assembly process have to be planned in advance, geometrical errors in part position and orientation as well as the manipulator tools and the manipulator itself have to be eliminated before robotic assembly may take place, which is very costly in general, [WHITNEY 89].

Achieving a more flexible handling of assembly processes, including dealing with randomly positioned and oriented or even moving parts requires dealing with problems resulting from higher uncertainty. This is the domain for new generations of vision based robotic manipulators.

Another area where visually controlled robotic manipulators may be advantageously used relates to applications in outer space, satellite retrieval for example, [KRISHEN et al.87], [HIRZINGER 87]. A typical scenario describes an unmaneuverable translat-ing and rotating satellite that should be retrieved by a (fairly large) robotic manipulator. In case that the manipulator is equipped with 3D-vision sensors and appropriate control strategies that cope with the a priori unkown movements of the satellite, this task can be done with some degree of autonomy.

The purpose of this chapter is to attack the general problem of a vision based position/orientation control (called "pose control" in the sequel) of a manipulator hand relative to an object that is to be grasped or manipulated (called "workpiece" in the sequel), which may be moving w.r.t. an initial coordinate frame. Force control, which is necessary during the contact phases of a manipulation process, see chapter 2, is not considered here.

Suggesting a concept for a visual pose control scheme one is faced with the following questions:

1.   How will relative movements of the robot/environment system be represented ?

2.   What kind of vision sensors should be used ? Where are they placed ?

3.   What kind of visual features should be used and how are they evaluated in order to get a dynamic description of relative movements ?

4.   How will nonlinear manipulator kinematics and dynamics be taken into ac-count ?

5.   What is the structure of a "visual controller" that processes descriptions of rela-tive movements and generates appropriate controls for the manipulator in order to guarantee a reasonable control behaviour ?

In this enumeration only the control-theoretical aspects of visual pose control are considered which are the topic of this work. Not considered are issues related to "higher visual skills" like object recognition, shape description, etc. and "reasoning" that means task level planning, planning of a stable grasp, etc., see [REMBOLD 88], [LOZANO, PEREZ, TAYLOR 89].

Before we present our concept and algorithms, a short review of existing approaches to the problems formulated above is desirable. [IKEUCHI et al. 84] and [INOUE, INABA 84] employed monocular vision with a camera mounted in a stationary relation to the base of a robot and suggested a control on a " look - then move - basis". Thus, they did not realize a true closed loop control, since the actions "sensing" and "moving" are performed sequentially. Furthermore, placing the visual sensors stationary with respect to the workspace of the manipulator, poses two problems:

- occlusions of workpieces due to motion of the manipulator can not be avoided in general,

- the inaccuracies of visual measurements increase with the distance between the camera and the workpiece, thus rendering the realization of accurate fine motions difficult.

These problems may be alleviated by mounting the visual sensors directly on the last link of the manipulator (eye - in hand technique) [KING 87], [HIRZINGER 87], [WEISS et al. 87].

[WEISS et al. 87] suggested a scheme consisting of a manipulator equipped with one camera mounted directly on its endeffector. The approach relies on a local transformation, that defines the relationship between velocities of certain features in the image plane and velocities of the endeffector, represented by a jacobian matrix. Starting from a linearization of the manipulator dynamics, model reference adaptive control is applied to each link of the manipulator independently. The authors report successful simulations for two and three degrees of freedom (DOF) examples, however no stability of their control scheme can be guaranteed and no moving workpieces have been considered.

A similar concept has been proposed by [RIVES et al. 90]. They stated the visual servo problem in very general terms including moving workpieces but then treated only the case of stationary workpieces.

[FÄSSLER et al. 90] realized a robot ping-pong player relying on measurements of a stationary stereo - camera system. Since the ping-pong ball is a point - like object its orientation is not of interest, only its position with respect to a world coordinate system is required. The manipulator dynamics has been particularly simple (two decoupled translational degrees of freedom).

To the authors knowledge there are no concepts for visual pose control strategies that explicitly take into account the highly complex structure of manipulator/environment systems including nonlinear manipulator dynamics and complex kinematics due to moving workpieces, yet guarantee stability and reasonable convergence.

The material presented in this chapter is a continuation of our earlier work, [GRAF-FUNDER, HARTMANN 90], [REN et al. 90].

4.2 Statement of the problem

We consider the general situation of an n - DOF manipulator ($n \geq 6$) moving relative to a workpiece in order to grasp or manipulate it, see fig. 4-1.

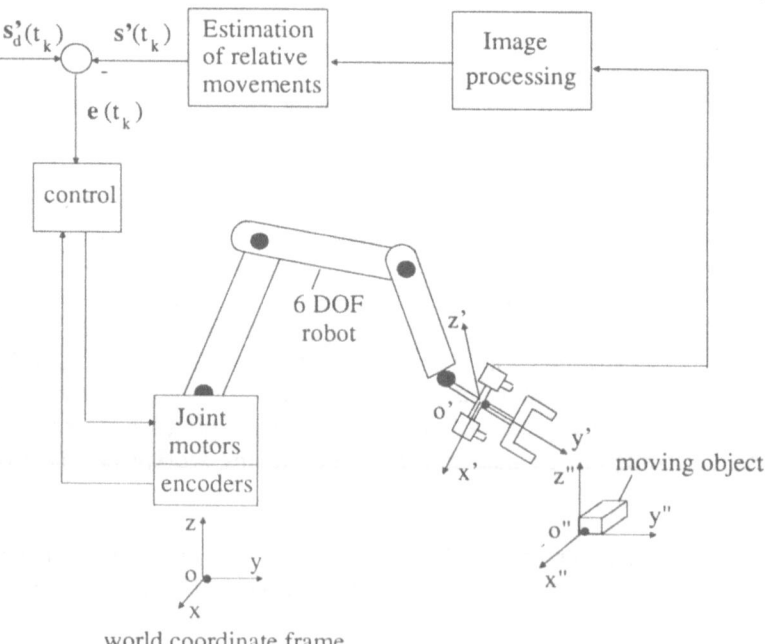

Fig. 4-1 Structure of visually controlled manipulator with
stereo cameras mounted on the endeffector

With regard to potential applications, named above, the workpiece is allowed to move in a 3D-space with arbitrary translations and rotations that are assumed not to be known a priori.

Visual sensing is provided by two cameras mounted on the endeffector of the manipulator, forming a stereosystem, see chapter 3.

Thus, we have exactly the situation treated in chap. 3 and we may use the same notation as has been introduced there. Hence,

- (O, x, y, z) denotes the world coordinate frame rigidly connected with the base of the manipulator.

- (O', x',y', z') denotes the endeffector coordinate frame which for simplicity coincides with the coordinate frame of the stereosystem (in chapter 3, this has been called "observer coordinate frame").

- (O", x", y", z") denotes the object coordinate frame which is rigidly connected to the object and it is assumed that all relevant geometric and kinematic properties of the object are defined w.r.t. this frame.

As stated in chap. 2 the manipulator dynamics is in general adequately represented by the usual nonlinear coupled system of differential equations

$$\underline{M}(q)\,\ddot{q} + \underline{C}(q,\dot{q}) + g(q) = \underline{\tau} \qquad (4\text{-}1)$$

According to the concepts which have been introduced in chap. 3, we may represent the relative position of the object w.r.t. the endeffector frame by the vector $\underline{r}'_{o''}$ pointing from the origin of the endeffector to the origin of the object coordinate frame. The alignment matrix $\tilde{R}$, which relates the two coordinate frames, represents the relative orientation of the object w.r.t. the endeffector. In chap. 3 an equivalent representation, by unit quaternions

$$\overline{\kappa}^T = [\,\sin(\frac{\theta}{2})\cdot n_x\,,\,\sin(\frac{\theta}{2})\cdot n_y\,,\,\sin(\frac{\theta}{2})\cdot n_z\,,\,\cos(\frac{\theta}{2})\,] = [\,\underline{\kappa}_v^T\,,\,\kappa_s\,] \qquad (4\text{-}2)$$

has been introduced, where $\underline{n}^T = [\,n_x\,,\,n_y\,,\,n_z\,]$ is a unit vector in the direction of the rotation axis, θ denotes the angle of rotation, $\underline{\kappa}_v$ denotes the vector part of the quaternion and κ_s its scalar part.

As has been explained in chapter 3 this representation is unique in case that $-\pi < \theta < \pi$; this is assumed in the sequel. Since the vector part $\underline{\kappa}_v$ suffices to define relative orientation, see chap. 3, the relative pose of the object w.r.t. the endeffector can be represented by

$$\underline{s}' = \begin{pmatrix} \underline{r'}_{o''} \\ \underline{\kappa}_v \end{pmatrix} \tag{4-3}$$

We further assume, that enough feature points and straight line segments can be extracted from the stereo-images and tracked over time so that estimations of the relative pose $\underline{s}'$ can be provided by two extended Kalman filters, as suggested in chapter 3.

Now, given a certain manipulation task, it is natural to define the task directly w.r.t. endeffector coordinates, i.e. we may want the endeffector to move according to an associated desired trajectory of the relative pose, called $\underline{s}'_d(t)$. Thus, the task of a relative movement control scheme is to drive the joint motors of the manipulator in such a way that the relative pose error

$$\underline{e}(t) := \underline{s}'_d - \underline{s}'(t) \tag{4-4}$$

converges to zero with a reasonable (prespecified) rate of convergence.

4.3 Theory of relative pose control

In order to derive an analytically motivated strategy for relative pose control, the dynamics of relative movements have to be considered.

Applying the same notation as in chap. 3 , let

$\underline{R}_c$ denote the alignment matrix expressing the relation between endeffector- and world coordinate frame,

$\underline{r}_{o'}$ denote the origin of the endeffector frame

$\underline{v}'_{o'} = \underline{R}_c \, \underline{\dot{r}}_{o'}$ denote the translational velocity of the endeffector expressed in coordinates of the endeffector frame,

$\underline{v}'_{o''} = \underline{R}_c \, \underline{\dot{r}}_{o''}$ denote the translational velocity of the object expressed in coordinates of the endeffector frame,

$\underline{\omega}_c$ denote the angular velocity of the endeffector with coordinates w.r.t. the world coordinate frame,

$\underline{\omega}'_b = \underline{R}_c \, \underline{\omega}_b$ denote the angular velocity of the object with coordinates w.r.t. the endeffector frame.

Recall from chap. 3 that the relative position $\underline{r}'_{o''}$ and the unit quaternion $\underline{\kappa}$ that represents the relative orientation obey the differential equations:

$$\dot{\underline{r}}'_{o''} = -\underline{\Omega}(\underline{\omega}_c)\,\underline{r}'_{o''} + \underline{v}'_{o''} - \underline{v}'_{o'} \tag{4-5}$$

$$\dot{\overline{\underline{\kappa}}} = \frac{1}{2}\underline{\Gamma}(\tilde{\underline{\omega}})\cdot\overline{\underline{\kappa}} \tag{4-6}$$

where

$$\underline{\Omega}(\underline{\omega}_c) = \begin{pmatrix} 0 & -\omega_{cz} & \omega_{cy} \\ \omega_{cz} & 0 & -\omega_{cx} \\ -\omega_{cy} & \omega_{cx} & 0 \end{pmatrix}$$

$$\underline{\Gamma}(\tilde{\underline{\omega}}) = \begin{pmatrix} 0 & \tilde{\omega}_z & -\tilde{\omega}_y & \tilde{\omega}_x \\ -\tilde{\omega}_z & 0 & \tilde{\omega}_x & \tilde{\omega}_y \\ \tilde{\omega}_y & -\tilde{\omega}_x & 0 & \tilde{\omega}_z \\ -\tilde{\omega}_x & -\tilde{\omega}_y & -\tilde{\omega}_z & 0 \end{pmatrix}$$

and $\tilde{\underline{\omega}} := \underline{\omega}'_b - \underline{\omega}_c$ denotes the relative angular velocity.

From (4-6) we find that the vector part of $\overline{\underline{\kappa}}$ obeys the differential equations

$$\dot{\underline{\kappa}}_v = -\frac{1}{2}\underline{\Omega}(\tilde{\underline{\omega}})\,\underline{\kappa}_v + \tilde{\underline{\omega}}\,\kappa_s \tag{4-7}$$

Since the antisymmetric matrix $\underline{\Omega}(\cdot)$ satisfies $\underline{\Omega}(\underline{a})\cdot\underline{b} = -\underline{\Omega}(\underline{b})\cdot\underline{a}$ for all $\underline{a},\underline{b}\in R^3$, (4-5) and (4-7) can be reformulated:

$$\dot{\underline{r}}'_{o''} = \underline{\Omega}(\underline{r}'_{o''})\,\underline{\omega}_c + \underline{v}'_{o''} - \underline{R}_c\,\dot{\underline{r}}'_{o'} \tag{4-8}$$

$$\begin{aligned}
\dot{\underline{\kappa}}_v &= \frac{1}{2}\underline{\Omega}(\underline{\kappa}_v)\,\tilde{\underline{\omega}} + \tilde{\underline{\omega}}\,\kappa_s \\
&= -\frac{1}{2}(\kappa_s\underline{I} + \underline{\Omega}(\underline{\kappa}_v))\,\underline{\omega}_c + \frac{1}{2}(\kappa_s\underline{I} + \underline{\Omega}(\underline{\kappa}_v))\,\underline{\omega}'_b
\end{aligned} \tag{4-9}$$

$\underline{I}$ denoting the 3x3 identity matrix.

The translational and angular velocities of the endeffector represented in base coordinates ($\dot{\underline{r}}_o$ and $\underline{\omega}_c$) are linked to the joint velocities by the manipulator jacobian, see [FU et al. 87]. For a six DOF robot this relationship takes the form, (chap. 2, eq. (2-13))

$$\begin{pmatrix} \dot{r}_{o'} \\ \omega_c \end{pmatrix} = \underline{J}(q) \cdot \dot{q} = [\underline{j}_1(q), \dots, \underline{j}_6(q)] \cdot \dot{q} \tag{4-10}$$

There are several ways to calculate the vectors $\underline{j}_i(q)$ either in closed form or numerically. One of the methods that yield closed form expressions is that of [WHITNEY 72], see also [FU et al. 87]. There it is shown that

$$\underline{j}_i(q) = \begin{cases} \begin{pmatrix} \dfrac{\Omega(z_{i-1}) \cdot {}^{i-1}p_6}{z_{i-1}} \end{pmatrix} & \text{if } joint\ i\ is\ rotational \\[2em] \begin{pmatrix} z_{i-1} \\ 0 \end{pmatrix} & \text{if } joint\ i\ is\ translational \end{cases}$$

where z_{i-1} is a unit vector pointing in the direction of motion of joint i expressed in base coordinates (for rotational joints this is the direction of the rotation axis) and ${}^{i-1}p_6$ is a vector pointing from the origin of the coordinate frame of the (i-1)-th joint to the origin of the endeffector coordinate frame expressed in coordinates w.r.t the base coordinate frame.

Substituting a rearranged version of (4-10)

$$\begin{pmatrix} \dot{r}_{o'} \\ \omega_c \end{pmatrix} = \begin{pmatrix} J_1(q) \\ J_2(q) \end{pmatrix} \cdot \dot{q} \tag{4-11}$$

into (4-8) and (4-9) gives differential equations that describe the impact of the joint positions and - velocities on the relative movements:

$$\dot{r}'_{o''} = [\Omega(r'_{o''})\, \underline{J}_2(q) - \underline{R}_c(q) \cdot J_1(q)]\dot{q} + \underline{v}'_{o''} \tag{4-14}$$

$$\dot{\underline{\kappa}}_v = -\frac{1}{2}(\kappa_s \underline{I} + \Omega(\underline{\kappa}_v))J_2(q) \cdot \dot{q} + \frac{1}{2}(\kappa_s \underline{I} + \Omega(\underline{\kappa}_v))\omega'_b \tag{4-15}$$

Given the desired trajectory of the relative pose $\underline{s}'_d$, then the error $\underline{e} = \underline{s}'_d - \underline{s}'$ obeys the differential equation:

$$\dot{\underline{e}}(t) = \dot{\underline{s}}'_d(t) - \begin{pmatrix} \Omega(r'_{o''})\, \underline{J}_2(q) - \underline{R}_c(q) \cdot J_1(q) \\[1em] -\dfrac{1}{2}(\kappa_s \underline{I} + \Omega(\underline{\kappa}_v))\underline{J}_2(q) \end{pmatrix} \dot{q} -$$

$$-\begin{pmatrix} \underline{v}'_{o''} \\ \frac{1}{2}\,(\,\kappa_s\,\underline{I}+\,\underline{\Omega}\,(\,\underline{\kappa}_v)\,)\,\underline{\omega}'_b \end{pmatrix} \tag{4-16}$$

or, written more compactly

$$\underline{\dot{e}}(t) = \underline{l}\,(\,\underline{\dot{s}}'_d,\overline{\underline{\kappa}},\underline{v}'_{o''},\underline{\omega}'_b\,) - \underline{L}\,(\,q\,,\underline{r}'_{o''}\,,\overline{\underline{\kappa}}\,)\,\dot{q} \tag{4-17}$$

where

$$\underline{l}\,(\,\underline{\dot{s}}'_d,\overline{\underline{\kappa}},\underline{v}'_{o''},\underline{\omega}'_b\,) := \underline{\dot{s}}'_d - \begin{pmatrix} \underline{v}'_{o''} \\ \frac{1}{2}\,(\,\kappa_s\,\underline{I}+\underline{\Omega}\,(\,\underline{\kappa}_v)\,)\,\underline{\omega}'_b \end{pmatrix}$$

$$\underline{L}\,(\,q\,,\underline{r}'_{o''}\,,\overline{\underline{\kappa}}\,) := \begin{pmatrix} \underline{\Omega}\,(\,\underline{r}'_{o''})\,\underline{J}_2(q)\ -\ \underline{R}_c\,(q)\cdot\underline{J}_1(q) \\ -\frac{1}{2}\,(\,\kappa_s\,\underline{I}+\,\underline{\Omega}\,(\,\underline{\kappa}_v)\,)\,\underline{J}_2\,(q) \end{pmatrix}$$

Looking at (4-16) reveals the highly nonlinear dependencies of the relative pose-error on the relative pose ($\underline{r}'_{o''}$ and $\overline{\underline{\kappa}}$), the velocities of the workpiece ($\underline{v}'_{o''}$ and $\underline{\omega}'_b$) and the actual state of the manipulator (q and $\dot{q}$).

Now, solving the problem of relative movement control means achieving asymptotic stability of the error differential equation (4-17) with an appropriate rate of convergence.

A possible requirement could be

$$\underline{\dot{e}}\,(t) := \underline{P}\,\underline{e}\,(t) \tag{4-18}$$

where all eigenvalues of $\underline{P}$ have negative real parts, guaranteeing exponential convergence of the relative pose error.

Equating (4-17) and (4-18) and solving for $\dot{q}$ gives the desired joint velocities

$$\dot{q}_d = \underline{L}^{-1}(\,q\,,\underline{r}'_{o''}\,,\overline{\underline{\kappa}}\,)\,[\,\underline{l}\,(\,\underline{\dot{s}}'_d,\overline{\underline{\kappa}},\underline{v}'_{o''}\,,\underline{\omega}'_b\,) - \underline{P}\,\underline{e}\ ] \tag{4-19}$$

The existence of the inverse of $\underline{L}$ is guaranteed if the jacobian of the manipulator is not singular (see also the remarks in chapter 2 ,sec. 3) and $\kappa_s = \cos(\frac{\theta}{2}) \neq 0$, (this is true, since we have assumed that $-\pi < \theta < \pi$) .

The relation (4-19) suggests that after having replaced the arguments of $\underline{l}$ and $\underline{L}$ by their respective estimates the desired joint velocities may be computed defining a nonlinear algebraic controller for the relative kinematics.

Conceptually, the output $\dot{q}_d$ may be fed into an underlying control loop for the joint velocities $\dot{q}$ that realizes approximately

$$\dot{q}(t) = \dot{q}_d(t) \tag{4-20}$$

Neglecting modeling- and measurement/estimation errors, and assuming that (4-20) holds exactly, the requirement (4-18) would be satisfied.

However, since the manipulator dynamics is highly nonlinear and coupled and the input signal energy is limited, the idealistic assumption (4-20) is not realistic, except in cases of very slow movements.

Therefore, the movement controller (4-19) must be modified appropriately in order to account for "non zero dynamics" of the underlying joint velocity control.

Control of joint velocities

In chap. 2 it has been shown, that for an n - DOF manipulator with nonlinear dynamics (4-1), a nonlinear decoupling contoller can be constructed which transforms the associated nonlinear state space model (2-3) into n linear time - invariant decoupled second order systems, one for each joint, see eq. (2-11). Adding input forces / moments $\underline{\tau}_c = \underline{M} \cdot \underline{w}$ to the decoupling forces / moments $\underline{\tau}_d$, the linear system (2-12) results :

$$\begin{pmatrix} \dot{z}_1 \\ \dot{z}_2 \\ \cdot \\ \cdot \\ \cdot \\ \dot{z}_n \end{pmatrix} = \begin{pmatrix} A_1 & 0 & \cdot & \cdot & 0 \\ 0 & A_2 & 0 & \cdot & \cdot & 0 \\ \cdot & & 0 & & & \cdot \\ \cdot & & & & & \cdot \\ \cdot & & & & & \cdot \\ 0 & 0 & \cdot & \cdot & \cdot & A_n \end{pmatrix} \begin{pmatrix} z_1 \\ z_2 \\ \cdot \\ \cdot \\ \cdot \\ z_n \end{pmatrix} + \begin{pmatrix} w_1 \\ w_2 \\ \cdot \\ \cdot \\ \cdot \\ w_n \end{pmatrix} \tag{4-21}$$

where z_i denote the position and velocity of the i-th joint, $\underline{w}^T = [\underline{w}_1{}^T, \cdots, \underline{w}_n{}^T]$ are the control - inputs and the matrices A_i may be chosen in order to assign the eigenvalues s_{i1}, s_{i2} of each subsystem according to a desired specification, provided that Re $[s_{i1}]$, Re $[s_{i2}] < 0$ for all i.

In case that the inputs $\underline{w}_i$ are chosen to be

$$\underline{w}_i{}^T = [\,0\,, w_i\,]$$

(4-22)

and defining the joint velocities $\dot{q}_i$ as outputs, the transfer-function of the i-th subsystem becomes:

$$G_i(s) = \frac{L[\,\dot{q}_i(t)\,]}{L[\,w_i(t)\,]} = \frac{s}{(s - s_{i1})\cdot(s - s_{i2})} \quad ,$$

(4-23)

$L[\,\cdot\,]$ denoting the Laplace - transform.

In order to control the joint velocities according to certain references $\dot{q}_d(t)$, linear velocity controllers of the form

$$G_{ci}(s) = K_i\,\frac{(s - s_{i1})\cdot(s - s_{i2})}{s^2}$$

(4-24)

may be applied. This results in closed loop transfer-functions

$$T_i(s) = \frac{L[\,\dot{q}_i(t)\,]}{L[\,\dot{q}_{di}(t)\,]} = \frac{K_i}{(s + K_i)}$$

(4-25)

Thus, the underlying joint velocity control system constitutes of n parallel decoupled linear control loops.

Modified movement controller

From (4-25) we conclude that

$$\dot{q}_i = \dot{q}_{di} - \ddot{q}_i\,/\,K_i$$

(4-26)

holds.

The difference between the true joint-velocities $\dot{q}_i$ and the inputs $\dot{q}_{di}$ is proportional to the joint-accelerations and decrease as the bandwidths of the joint-velocity control loops increase.

Clearly, there are bounds on the available joint-forces/torques that limit the realizable bandwidths so that the terms $\ddot{q}_i/K_i$ may not be negligible.

The relations (4-26) may be rewritten in vector form

$$\dot{q} = \dot{q}_d - \underline{D}^{-1} \cdot \ddot{q} \quad ; \quad \underline{D} = diag\,[\,K_1\,, \cdots, K_n\,] \tag{4-27}$$

Substitution into (4-17), equating the result with the requirement (4-18) and solving for $\dot{q}_d$ defines the modified nonlinear control law:

$$\dot{q}_d = \underline{L}^{-1}\,(\,q\,,\underline{r}'_{o''}\,,\overline{\underline{\kappa}}\,)\cdot[\,\underline{l}(\,\dot{\underline{s}}'_d\,,\overline{\underline{\kappa}}\,,\underline{v}'_{o''}\,,\underline{\omega}_b'\,) - \underline{P}\cdot\underline{e}\,] + \underline{D}^{-1}\cdot\ddot{q} \tag{4-28}$$

Discretization

In order to get realistic implementations of the presented theory the decoupling controller and the movement control algorithm (4-28) have to be realized in discrete time.

The nonlinear decoupling may be realized using for example a recursive Newton-Euler algorithm to calculate the necessary decoupling forces / moments $\underline{\tau}_d$, chap. 2, eq. (2-9). It requires $111n - 4$ additions and $132n$ multiplications for an n-DOF manipulator, see e.g. [Fu et al. 87]. With n = 6 and using parallel processing, sampling rates on the order of T =10ms can be achieved. Errors due to discrete time processing and inaccuracies in model building are at least partially compensated by the velocity controllers (4-24), which may operate in continous time.

The movement control algorithm (4-28) requires the manipulator-jacobian $\underline{J}(q)$,see (4-10), at each sampling instant. In case that a recursive Newton - Euler algorithm is chosen to compute the decoupling forces / moments, the "strobing technique" suggested by [FU el al. 87] may be used in order to determine the elements of the jacobian numerically. Since results already computed by the Newton - Euler algorithm can be exploited, these computations add only little to the overall computational burden.

Furthermore, the movement control algorithm requires knowledge of the kinematic quantities that define the state of the relative movements namely, $\underline{r}'_{o''}$, $\overline{\underline{\kappa}}$, $\underline{v}'_{o''}$ and $\underline{\omega}'_b$.

Estimates of these are available from the extended Kalman filters for estimation of the relative motion states, see chap. 3. Between two correction-steps of the filters, the estimates at $t = t_{k-1}$ are propagated by integrating the state equations describing the relative movements, yielding $\hat{\underline{x}}\,(\,\tau/t_{k-1}\,)$, $t_{k-1} \le \tau < t_k$ (eq. (3-50) for the rotational motion states, similar relations hold for the translational motion states). Splitting the interval $[\,t_{k-1}\,,\,t_k\,)$ into smaller sub-intervals of length T_2, estimates for the motion states become available from the filter propagations at the ends of each sub-interval. Thus, although the stereo-processing may be quite time consuming and therefore the intervals between two filter updates $[\,t_{k-1}\,,\,t_k\,)$ can be relatively long, the movement-control algorithm can operate using the propagated estimates according to a higher sampling rate $1/T_2$.

In the sequel discrete versions of the movement-controller will be derived, assuming a constant sampling rate $1/T_2$. Substituting (4-27) into the error differential equation (4-17) gives

$$\dot{\underline{e}}(t) = \underline{l}\,(\,\dot{\underline{s}}_d{}'(t)\,,\overline{\underline{\kappa}}(t)\,,\underline{v}'_{o\cdot\cdot}(t)\,,\underline{\omega}_b{}'(t)\,)\,-$$

$$-\,\underline{L}\,(\,\underline{q}(t)\,,\underline{r}'_{o\cdot\cdot}(t)\,,\overline{\underline{\kappa}}(t)\,)\cdot(\,\dot{\underline{q}}_d\,(t)-\underline{D}^{-1}\cdot\ddot{\underline{q}}(t)\,) \qquad (4\text{-}29)$$

Holding $\dot{\underline{q}}_d(t)$ constant between the sampling instants $t_k = k\cdot T_2$ (according to a zero-order hold) and integrating (4-29) over the sampling interval gives

$$\underline{e}((k{+}1)T_2) = \underline{e}(kT_2) + \int_{kT_2}^{(k+1)T_2}\underline{l}(\tau)\,d\tau \;-\; \int_{kT_2}^{(k+1)T_2}\underline{L}(\tau)\,d\tau\cdot\dot{\underline{q}}_d(kT_2)$$

$$+ \int_{kT_2}^{(k+1)T_2}\underline{L}(\tau)\cdot\underline{D}^{-1}\cdot\ddot{\underline{q}}(\tau)\,d\tau \qquad (4\text{-}30)$$

We assume that we may substitute $\underline{l}(\tau)$ and $\underline{L}(\tau)$ by the linear parts of their Taylor expansions at $\tau = kT_2$:

$$\underline{l}(\tau) \approx \underline{l}\,(kT_2) + \dot{\underline{l}}(kT_2)\cdot(\tau - kT_2) \qquad (4\text{-}31)$$

$$\underline{L}(\tau) \approx \underline{L}\,(kT_2) + \dot{\underline{L}}\,(kT_2)\cdot(\tau - kT_2) \qquad (4\text{-}32)$$

Then an evaluation of the integrals in (4-30) yields (the sampling period T_2 suppressed):

$$\int_{kT_2}^{(k+1)T_2}\underline{l}(\tau)\,d\tau = \underline{l}\,(k)\cdot T_2 + \dot{\underline{l}}\,(k)\cdot T_2^2/\,2 \qquad (4\text{-}33a)$$

$$\int_{kT_2}^{(k+1)T_2}\underline{L}(\tau)\,d\tau = \underline{L}\,(k)\cdot T_2 + \dot{\underline{L}}\,(k)\cdot T_2^2/\,2 \qquad (4\text{-}33b)$$

$$\int_{kT_2}^{(k+1)T_2}\underline{L}(\tau)\,\underline{D}^{-1}\,\ddot{\underline{q}}(\tau)\,d\tau = \underline{D}^{-1}\big\{\underline{L}\,(k)\cdot\Delta\,\dot{\underline{q}}(k) + \dot{\underline{L}}\,(k)\cdot[\,T_2\,\dot{\underline{q}}(k{+}1) - \Delta\,\underline{q}(k)\,]\big\} \qquad (4\text{-}33c)$$

where

$$\Delta \underline{q}(k) := \underline{q}(k+1) - \underline{q}(k)$$

$$\Delta \underline{\dot{q}}(k) := \underline{\dot{q}}(k+1) - \underline{\dot{q}}(k)$$

On the other hand, integrating (4-18) over one sampling interval gives

$$\underline{e}((k+1)T_2) = e^{\underline{P} \cdot T_2} \cdot \underline{e}(kT_2) := \underline{\Phi} \cdot \underline{e}(kT_2) \tag{4-34}$$

In case that the eigenvalues of P are given by $p_1, p_2, \cdots, p_n$ then the eigenvalues of $\underline{\Phi}$ are $e^{p_1 \cdot T_2}, \cdots, e^{p_n \cdot T_2}$.

A discrete version of the movement controller (4-28) results after substitution of (4-33) into (4-30), equating the right hand sides of (4-30), (4-34) and solving for $\underline{\dot{q}}_d(k)$:

$$\underline{\dot{q}}_d(k) = [\underline{L}(k) \cdot T_2 + \underline{\dot{L}}(k) \cdot T_2{}^2/2]^{-1} \cdot [[\underline{I} - \underline{\Phi}] \cdot \underline{e}(k) + \underline{l}(k) \cdot T_2 +$$

$$+ \underline{\dot{l}}(k) T_2^2 + \underline{D}^{-1} [\underline{L}(k) \cdot \Delta \underline{\dot{q}}(k) + \underline{\dot{L}}(k) \cdot [T_2 \cdot \underline{\dot{q}}(k+1) - \Delta \underline{q}(k)]]] \tag{4-35}$$

Here, $\underline{\dot{l}}(k)$ and $\underline{\dot{L}}(k)$ may be approximated by discrete differences

$$\underline{\dot{l}}(k) \approx 1/T_2 \cdot [\underline{l}(k+1) - \underline{l}(k)] =: \Delta \underline{l}(k) / T_2 \tag{4-36a}$$

$$\underline{\dot{L}}(k) \approx 1/T_2 \cdot [\underline{L}(k+1) - \underline{L}(k)] =: \Delta \underline{L}(k) / T_2 \tag{4-36b}$$

Calculating $\underline{l}(k+1)$ and $\underline{L}(k+1)$ needs the values of their arguments at time $t_{k+1} = (k+1) \cdot T_2$. Predictions for these are available from the movement filters, see chapter 3. The joint velocities $\underline{\dot{q}}(k), \underline{\dot{q}}(k+1)$ must be established either by an observer for the manipulator states or by simple numerical differentiation of the joint positions which are usually measured using high resolution - encoders with sufficiently small quantization noise.

A simpler version of the movement controller (4-35) results in case we neglect $\underline{\dot{l}}(k T_2)$ and $\underline{\dot{L}}(k T_2)$ in the Taylor expansions for $\underline{l}(\tau)$ and $\underline{L}(\tau)$ (4-31), (4-32). Choosing the sampling period T_2 and the eigenvalues of $\underline{P}$ such that

$$\underline{\Phi} = e^{\underline{P} \cdot T_2} \approx \underline{I} + \underline{P} \cdot T_2 \tag{4-37}$$

holds, then (4-35) simplifies to

$$\underline{\dot{q}}_d(k) = \underline{L}^{-1}(k) [-\underline{P} \cdot \underline{e}(k) + \underline{l}(k)] + \underline{D}^{-1} \cdot \Delta \underline{\dot{q}}(k) / T_2. \tag{4-38}$$

4.4 Results

For the purpose of easy illustration we present simulation results for a three DOF manipulator with three parallel rotational joints for which a dynamical model has been derived in chapter 2. The simulation set up is depicted in fig. 4-2.

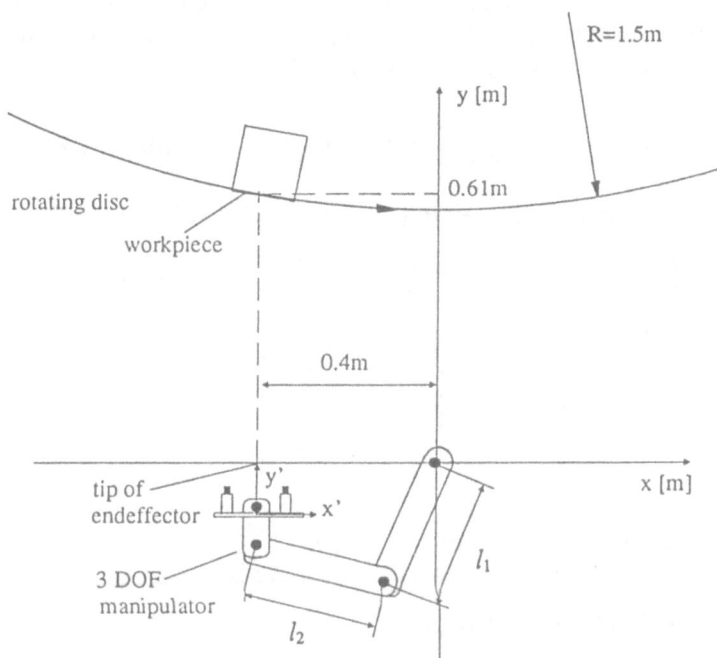

Fig. 4-2 Simulation set - up

The motions of the endeffector are restricted to the xy-plane having two translational DOF ($\underline{r}_{o'}^T = [\, x_{o'} , y_{o'}\,]$) and one DOF for orientation (angle φ).

The kinematic model in this case takes on the form, (see fig. 2-1)

$$\underline{r}_{o'} = \underline{T}_1(\underline{q}) = \begin{pmatrix} l_3 \cdot \sin(q_{123}) + l_2 \cdot \sin(q_{12}) + l_1 \cdot \sin(q_1) \\ -\, l_3 \cdot \cos(q_{123}) - l_2 \cdot \cos(q_{12}) - l_1 \cdot \cos(q_1) \end{pmatrix} \qquad (4\text{-}39)$$

$$\varphi = T_2(\underline{q}) = q_{123}$$

where $\underline{q}^T = [\,q_1\,,q_2\,,q_3\,]$, $q_{123} = q_1 + q_2 + q_3$, $q_{12} = q_1 + q_2$.

The relative pose in case of planar movements is defined to be

$$\underline{s}' = \begin{pmatrix} x'_{o''} \\ y'_{o''} \\ \kappa_{vz} \end{pmatrix} \tag{4-40}$$

where $\kappa_{vz} = \sin(\theta/2)$, and θ denotes the angle between endeffector and object coordinate frame. After having calculated the jacobians $\underline{J}_1(\underline{q})$ and $\underline{J}_2(\underline{q})$, the matrices $\underline{L}$ and $\underline{l}$ of the movement controller (4-28) can be determined:

$$\underline{L} = \begin{pmatrix} -y'_{o''} - s\theta{\cdot}j_{11} + c\theta{\cdot}j_{21} & -y'_{o''} - s\theta{\cdot}j_{12} + c\theta{\cdot}j_{22} & -y'_{o''} - s\theta{\cdot}j_{13} + c\theta{\cdot}j_{23} \\[2mm] x'_{o''} - c\theta{\cdot}j_{11} + s\theta{\cdot}j_{21} & x'_{o''} - c\theta{\cdot}j_{12} + s\theta{\cdot}j_{22} & x'_{o''} - c\theta{\cdot}j_{13} + s\theta{\cdot}j_{23} \\[2mm] -\dfrac{1}{2}\cdot\kappa_s & -\dfrac{1}{2}\cdot\kappa_s & -\dfrac{1}{2}\cdot\kappa_s \end{pmatrix} \tag{4-41}$$

$$\underline{l} = \begin{pmatrix} -v'_{o''} \\[2mm] -\dfrac{1}{2}\cdot\kappa_s\cdot\omega'_{bz} \end{pmatrix}$$

where $c\theta := \cos(\theta)$, $s\theta := \sin(\theta)$, $\underline{v}'_{o''} = \underline{R}_c\cdot\begin{pmatrix}\dot{x}_{o''} \\ \dot{y}_{o''}\end{pmatrix}$,

j_{ij} denote the elements of $\underline{J}_1(\underline{q})$.

It may be shown that $\underline{L}$ is non-singular provided that $\kappa_s = \cos(\theta/2) \neq 0$. This is the case, since we have assumed that $-\pi < \theta < \pi$.

The stereo-camera system is mounted 0.1m behind the tip of the endeffector, the distance between the cameras is 0.1m, their focal lengths are both 7.2mm.

We simulated a quadratic object lying on a disc of radius 1.5m which rotates with a constant angular velocity $\omega_{bz} = 0.22\ rad/s$.

The background of this simulation set up could be a subtask of an assembly- process where an assembly-manipulator has to fetch a workpiece from a rotating parts feeder. This means, in order to safely grasp the workpiece from the feeder the position and orientation of the endeffector (gripper) has to be synchronized with that of the moving workpiece during the approach phase of the fetch maneuver. The object-coordinate system is defined as depicted in fig. 4-3, hence a suitable reference for the desired relative pose is simply $\underline{s}'_d(t) \equiv [\,0\,,0.1\,,0\,]^T$.

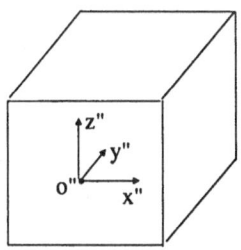

Fig 4-3 Definition of object coordinate frame

The movement control algorithm has been discretized according to the simple version (4-38) with a sampling interval $T_2 = 10\ ms$. The nonlinear decoupling is done every $T_1 = 10\ ms$.

Although the movements in this example are planar, the same estimation-process as has been described in chapter 3 was used in this simulation (3D-filters with updates calculated every 100 ms). The visual measurements (feature-points and line segments) were the same as those described in 3.7, with the exception, that all four corner-points of the object's front-face have been used in order to calculate estimations for the position of the object-origin (arithmetic mean of corner-point coordinates). Errors in the estimated quantities are due to quantization in the image planes of the stereo system (resolution 256×256). The movement filters need the transformed linear velocity of the endeffector origin $\underline{v}'_{o'} = \underline{R}_c \cdot \underline{\dot{r}}_{o'}$ (which is easily obtained by taking the time derivative of (4-39) and rotating the result by φ) and the angular velocity of the endeffector $\omega_c = \dot{\varphi} = \dot{q}_1 + \dot{q}_2 + \dot{q}_3$. The joint-positions and -velocities have been assumed known.

After the simulation is started, the manipulator rests at its initial position for a short time interval (0.5s) allowing the estimates for the relative pose and the velocities of the workpiece $\underline{v}'_{o''}$ and ω'_{bz} to converge.

In fig. 4-4 the position of the tip of the endeffector is shown, which converges smoothly to the trajectory of the workpiece.

The results for the relative position $x'_{o''}$, $y'_{o''}$, and the relative orientation κ_{vz} can be seen in fig. 4-5, 4-6 and 4-7 (full lines denoting true values, dotted lines denoting updated estimates) . After 0.5s, when the movement control starts, the relative pose converges nearly exponentially to the desired values.

Fig. 4-8, 4-9 and 4-10 show the results for the joint velocities. Here the full lines denote the desired values which were calculated by the movement controller (4-38) and the dotted lines denote the actual velocities.

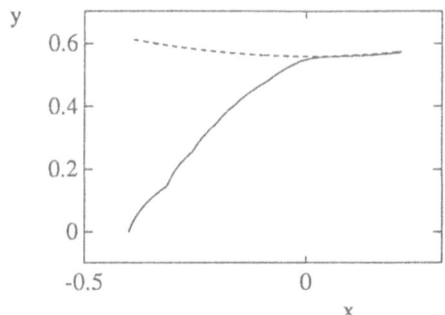

Fig. 4-4 Position of tip of endeffector (—)
and position of object origin (- -)
in the xy-plane

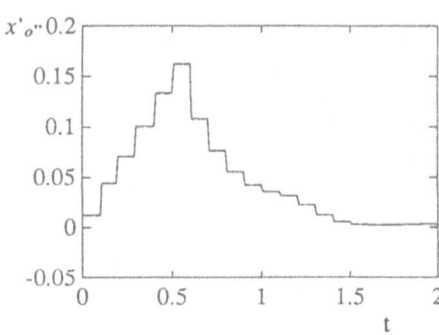

Fig. 4-5 Relative x-coordinate of object origin
(x'_{o}")

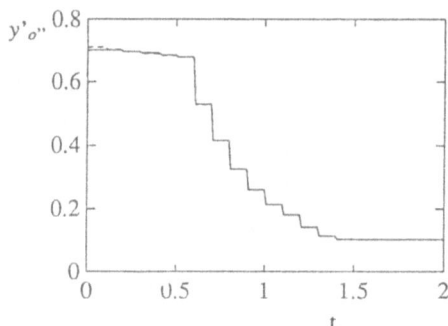

Fig. 4-6 Relative y-coordinate of object
(y'_{o}")

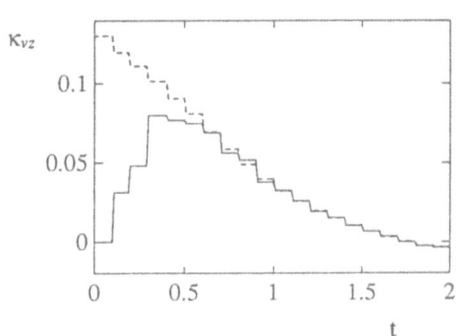

Fig. 4-7 Relative orientation (κ_{vz})

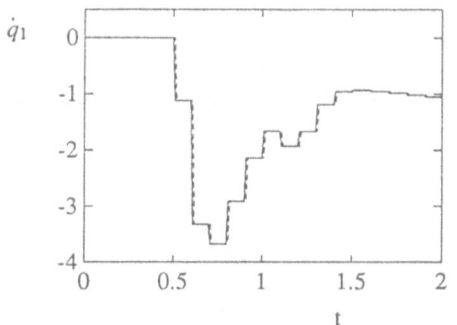

Fig. 4-8 Velocity of first joint ($\dot{q}_1$)

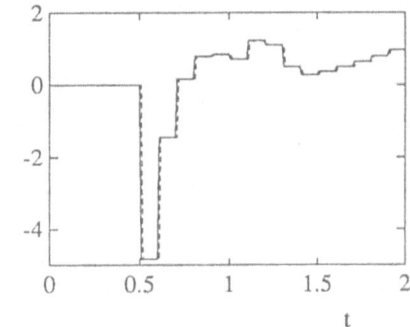

Fig. 4-9 Velocity of second joint ($\dot{q}_2$)

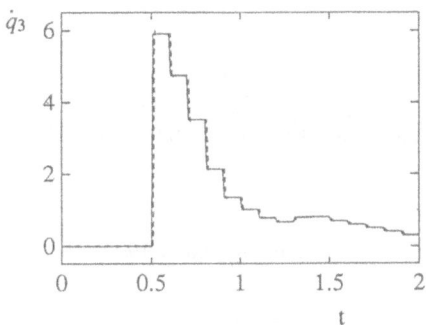

Fig. 4-10 Velocity of third joint ($\dot{q}_3$)

4.5 References

[FAESSLER et al. 90]
Faessler, H.; Beyer, H.A.; Wen, J.: " A robot ping pong player: optimized mechanics, high
performance 3D vision, and intelligent sensor control"
Robotersysteme, 6, 1990, pp. 161-170

[FU et al. 87]
Fu, K.S.; Gonzalez, R.C.; Lee, C.S.G.: "Robotics: Control, Sensing, and Intelligence"
McGraw Hill, 1987

[GAFFUNDER, HARTMANN 90]
Graffunder, A.; Hartmann, I.: "Towards Modelling Moving Objects Part One: Estimation of
3D-Relative Movements from Stereo-Image Sequences"
Preprints of the IFIP TC 7 Conference on Modelling the Innovation, Rome, March 1990, pp. 27-34

[HIRZINGER 87]
Hirzinger, G.:"The Space and Telerobotic Concepts of DFVLR ROTEX"
Proc. IEEE International Conference on Robotics and Automation, Raleigh, North Carolina, 1987,
pp. 138-150

[IKEUCHI et al. 84]
Ikeuchi, K.; Horn, B.K.P.; Nagata, S.; Callahan, T.; Feingold, O.: "Picking up an Object from a Pile
of Objects"
Robotics Research, The 1. International Symposium, M. Brady and R. Paul editors, 1984

[INOUE, INABA 84]
Inoue, H.; Inaba, M.: "Hand-Eye Coordination in Rope Handling"
Robotics Research, The 1. International Symposium, M. Brady and R. Paul editors, 1984

[KING et al. 87]
King, F.G.; Puskorius, G.V.; Yvan, F.; Meier, R.C.; Jeyabalan, V.; Feldkamp, L.A.: "Vision Guided
Robots for Automatic Assembly"

Proc. IEEE International Conference on Robotics and Automation, Philadelphia, Pennsylvania, 1988, pp. 1611-1616

[KRISHEN et al. 87]
Krishen, K.; Figueiredo, R.J.P.; Graham, O.: "Robotic Vision/Sensing for Space Applications"
Proc. IEEE International Conference on Robotics and Automation, Raleigh, North Carolina, 1987, pp. 138-150

[LOZANO-PEREZ 90]
Lozano-Perez, T.: Foreword
In : "Autonomous Robot Vehicles", Springer-Verlag, I.J. Cox, G.T. Wilfong editors, 1990

[LOZANO-PEREZ, TAYLOR 89]
Lozano-Perez, T.; Taylor, R.H.: "Geometric Issues in Planning Robot Tasks"
In : Robotics Science, MIT Press, Cambridge, Mass., M. Brady editor, 1989

[REMBOLD 88]
Rembold, I.U.: "The Karlsruhe Autonomous Mobile Assembly Robot"
Proc. IEEE International Conference on Robotics and Automation, Philadelphia, Pennsylvania, 1988, pp. 598-603

[REN et al. 90]
Ren, Z.; Graffunder, A.; Hartmann, I.: "Towards Manipulating Moving Objects Part Two: Intelligent Control of Robot-Endeffector Using Visual Feedback"
Proceedings of the IFIP TC 7 Conference on Modelling the Innovation, Rome, March 1990, North Holland, Elsevier Science Publishers, pp. 155-162

[RIVES et al. 90]
Rives, I.; Chaumette, F.; Espiau, B.: "Visual Servoing Based on a Task Function Approach"
Experimental Robotics I, The first Int. Symp. Montreal, June 19-21, 1989, Series Lecture Notes in Control and Information Sciences, No 139, Springer-Verlag, 1990

[WEISS et al. 87]
Weiss, L.E.; Sanderson, A.C.; Neuman, C.P.: "Dynamic Sensor-Based Control of Robots with Visual Feedback"
IEEE Journal of Robotics and Automation, Vol. RA-2, No.5, 1987, pp. 404-417

[WHITNEY 72]
Whittney, D.E.: "The Mathematics of Coordinated Control of Prosthetic Arms and Manipulators"
Trans. ASME Journal of Dynamic Systems, Measurement and Control, vol. 122, 1972, pp. 303-309

[WHITNEY 89]
Whittney, D.E.: "A Survey of Manipulation and Assembly: Development of the Field and Open Research Issues"
In Robotics Science, MIT Press, Cambridge, Mass., M. Brady editor, 1989

5 Endeffector Force Approximation

M. Boldin

5.1 Introduction

Endeffector forces play an important role in certain manufactoring tasks, such as screwing or inserting, in path tracking tasks, such as contour following or grinding, in tasks where e.g. two robots handle the same object and in mobile robots where the robot arm and consequently the end effector is influenced by platform movements. Before any force control scheme can be implemented there has of course to be a certain force measurement or sensing. In the literature there are mainly three approaches to force sensing. The first is the use of wrist force sensors, i.e. sensors consisting of mechanical elements equipped with strain gauges which convert displacement to force signals (see e.g. [SHIMANO, ROTH 79]). There are different designs available, but the principle remains the same. The advantage of these force sensors is the accuracy of the measurement and the fact that exact knowledge of the endeffector forces is obtained. Their disadvantage is the relatively high price, that makes them impossible to use when low cost solutions are requested. Another method is to measure or monitor the joint torques, which is either done by torque sensors mounted on each joint or by analyzing the actuator current [SHIMANO, ROTH 79], [NAGHDY et al. 85]. This was very popular for master-slave robots. By this method even forces that do not occur at the endeffector will be recognized and therefore situations where the arm hits obstacles will be realized. However there remains the costs for torque sensors. The third possibility encountered in the literature is the consideration of the robot as a constrained mechanical system and the use of descriptor systems theory [WANG, MCCLAM-ROCH 89], [HUANG, TZENG 89]. It has to be assumed that the constraints given by the contact surface and therefore the geometry of the contact surface itself is known, what can not be assured in e.g. contour following tasks.

In this article another method is considered. For linear systems the use of state observers and disturbance observers is quite common. Since the robot is a dynamical system and the endeffector forces can be interpreted as disturbance variables it seems quite useful to consider a disturbance observer to evaluate the endeffector forces. The problem that arises is the fact that the robot is a highly nonlinear coupled multivariable system and the design of nonlinear observers is not as simple as in the linear case. The following article is organized as follows: in 5.2 the possibility of using a disturbance observer for the endeffector forces is shown, in 5.3 a brief review of nonlinear observation methods is given and a special approximative technique is described, 5.4 shows the design for endeffector force observation and gives some examples. 5.5 draws the conclusions.

5.2 Observers for endeffector forces

In the following the possibility of observing endeffector forces is shown. The robot model excluding actuator dynamics and assuming friction is neglectable, can be written in state space form (see chap. 2, eq. (2-3))

$$\underline{x}_1 = \underline{q} \, , \underline{x}_2 = \underline{\dot{q}}$$

$$\underline{\dot{x}}_1 = \underline{x}_2$$

$$\underline{\dot{x}}_2 = \underline{M}^{-1}(\underline{x}_1)\left[\underline{\tau} - \underline{h}(\underline{x}_1, \underline{x}_2) - \underline{g}(\underline{x}_1) - \underline{J}^T \underline{F}_e \right] \qquad (5\text{-}1)$$

If the endeffector forces are represented by additional state variables $\underline{x}_3 = \underline{F}_e$ an augmented state space model has the form

$$\underline{\dot{x}}_1 = \underline{x}_2$$

$$\underline{\dot{x}}_2 = \underline{M}^{-1}(\underline{x}_1)\left[\underline{\tau} - \underline{h}(\underline{x}_1, \underline{x}_2) - \underline{g}(\underline{x}_1) - \underline{J}^T \underline{x}_3 \right]$$

$$\underline{\dot{x}}_3 = \underline{0}$$

If this model is observable with respect to some measured variable, e.g. joint positions $\underline{y}(t) = \underline{q}(t)$ and possibly joint velocities $\underline{y}(t) = [\underline{q}(t)\ \underline{\dot{q}}(t)]$, an observer for this model would reconstruct the endeffector forces. The observability will be checked by a special observability criterion in 5.4. If any a priori knowledge of the occuring force is at hand, it can be considered in the state space model by another right hand side of $\underline{\dot{x}}_3$.

5.3 Nonlinear state space observation

The nonlinear observation problem is not yet completely solved as the linear case is. There are several approaches that shall be briefly mentioned. The most popular method is based on nonlinear transformations, slightly differently formulated by the authors, that converts the nonlinear system to be observed in a special canonical form (see e.g. [BESTLE, ZEITZ 83], [KRENER, RESPONDEK 85], [WALCOTT et al. 87]. For this canonical form the observer design is quite simple and through back-transformation the observer is obtained in original coordinates. The problem of nonlinear observer design is thus transferred to the determination of the nonlinear transformation, which can be a hideous task or even more crucial, which can not be solved. So however the technique is interesting mathematically it is quite problematic from the practical point of view. Other methods that are based e.g. on Ljapunov theory are also restricted to special system classes or require further investigations (see e.g. [KOU et al. 75], [HARTMANN, LANDGRAF 89]. If analytically exact observer design is as complicated it seems useful to consider approximations. An approximation to nonlinear observers was given by [SCHÖNWANDT 73], that will be reviewed in the following.

The measured variable $\underline{y}(t)$ of a nonlinear dynamic system

$$\underline{\dot{x}} = f(\underline{x}, \underline{u}),$$

$$\underline{y} = g(\underline{x}),$$

$$\underline{x}(0) = \underline{x}_0 \in S, t \in [0, T],$$

$$\underline{x}(t) \in R^n, \underline{u}(t) \in R^r, \underline{y}(t) \in R^p, f : R^n \times R^r \to R^n, g : R^n \to R^p, S \subset R^n,$$

having only analytic nonlinearities can be expanded

$$\underline{y}(t) = \sum_{i=0}^{\infty} \underline{y}_i \, t^i, t \in [0,T],$$

where

$$\underline{y}_i = \frac{1}{i!} \frac{d^i \underline{y}}{d t^i} \Big|_{t=0}.$$

If the infinitly many coefficients of this series are written in vector form

$$\{\underline{y}_i\} = \{\underline{y}_1, \underline{y}_2, \dots \},$$

an operator notation for the relation between the initial condition $\underline{x}(0)$ and the coefficients can be chosen

$$\{\underline{y}_i\} = K_u \, \underline{x}_0,$$

where K_u is an operator mapping for some special $\underline{u}(t)$ from the space S of possible initial states to a space of sequences of vectors and an observability criterion can be formulated:

The system is observable if and only if the mapping K_u is one-to-one on S.

Thus the initial system state could be evaluated if the coefficients of the expansion were known and K_u^{-1} can be evaluated. For systems with analytic nonlinearities this can be generalized in the form, that the system state at time instant t_i can be evaluated knowing K_u^{-1} and the coefficients of the expansion for $t = t_i$. To avoid direct derivations of the measured variable Schönwandt shows that the coefficients y_i can be approximated using orthonormal projections.

Considering only one component of the output variable $y_k(t)$, this analytic time function (it is in fact analytic since the system contains only analytic nonlinearities) can be expanded using any other orthonormal basis of the space of square integrable functions $L_2 [0, T]$

$$y_k(t) = \sum_{n=0}^{\infty} \gamma_{kn} P_n(t)$$

and

$$\hat{y}_k(t) = \sum_{n=0}^{N} \gamma_{kn} P_n(t)$$

is an approximation in the least-square sense. If polynomials (e.g. Legendre polynomials)

$$P_n(t) = \sum_{q=0}^{n} \alpha_{nq} t^q$$

are used, the approximation can be rewritten

$$\hat{y}_k(t) = \sum_{i=0}^{N} s_{ki} t^i$$

with

$$s_{ki} = \sum_{j=i}^{N} \gamma_{kj} \alpha_{ji}.$$

Schönwandt shows that this approximation s_{ki} converges uniformly to y_{ki} if N goes to infinity. He then gives a possibility to evaluate the s_{ki} by a chain of integrations. Defining

$$\underline{A} = \begin{pmatrix} \alpha_{00} & 0 & 0 & \dots & 0 \\ \alpha_{10} & \alpha_{11} & 0 & \dots & 0 \\ \dots & \dots & \dots & \dots & \dots \\ \alpha_{N0} & \alpha_{N1} & \dots & \dots & \alpha_{NN} \end{pmatrix},$$

$$\underline{C} = \begin{pmatrix} 1 & 0 & \dots & 0 \\ 0 & -1 & \dots & \dots \\ \dots & \dots & \dots & \dots \\ 0 & 0 & \dots & (-1)^N \end{pmatrix},$$

$$\underline{B}_k = \begin{pmatrix} \beta_{k0} & \beta_{k1} & \dots & \beta_{kN} \end{pmatrix}^T,$$

the vector

$$\underline{S}_k = \begin{pmatrix} s_{k0} & s_{k1} & \dots & s_{kN} \end{pmatrix}$$

$$= \underline{C}\,\underline{A}^T\,\underline{A}\,\underline{B}_k$$

can be evaluated, where α_{ij} are the coefficients of the polynomials as above and

$$\beta_{kq} = q! \int_0^{T\sigma_q} \int_0^{\sigma_1} \dots \int_0^{\sigma_1} y_k(\sigma_0)\, d\sigma_0 \dots d\tau$$

is the output of a chain of integrators.

Thus an approximation to an observer can be built up as in fig. 5-1.

A block called analyser gives approximations of the derivatives of the output variable $\hat{y}(k\,T)$, $k=1, 2, \dots$ with the period T, that are needed to evaluate estimates of the state variables $\hat{x}(k\,T)$, $k=1, 2, \dots$. It should be clear to see, that the accuracy strongly depends on the order of the polynomial approximation and on the integration interval. Choosing the right values can only be done by careful simulation.

For a more detailed description of the algorithm see the original paper [SCHÖN-WANDT 73].

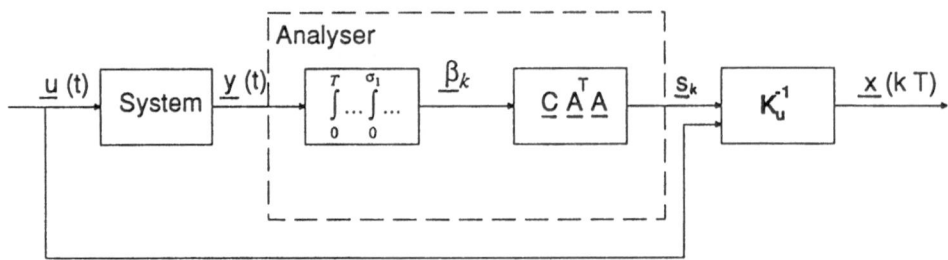

Fig. 5-1 Approximative State Observer

5.4 Approximative endeffector force observation

Having the knowledge of the approximative observer design of Schönwandt, an approximative observer for the endeffector forces will be developed. Using the model that was given in 5.2 (eq. (5-1)) it can first be verified, that the augmented state space model of the robot is observable in the sense of Schönwandt. Measuring the joint positions $\underline{y}(t) = \underline{q}(t)$, with $\underline{y}$ and $\underline{q}$ nx1-vectors, if an n degree of freedom manipulator is taken, and considering their derivatives $\underline{\dot{y}}(t)$ and $\underline{\ddot{y}}(t)$, the observability with respect to the criterion of Schönwandt can be verified by:

$$\begin{pmatrix} \underline{x_1} \\ \underline{x_2} \\ \underline{x_3} \end{pmatrix} = \begin{pmatrix} \underline{y} \\ \underline{\dot{y}} \\ \underline{J}^{T}[\underline{\tau} - \underline{M}\,\underline{\ddot{y}} - \underline{h}(\underline{y}, \underline{\dot{y}}) - \underline{g}(\underline{y})] \end{pmatrix}.$$

Here the right hand side is the operator K_u^{-1}, if $\underline{\tau}$ is understood as the input variable $\underline{u}$. Thus the measured joint positions $\underline{y}(t)$ have to be processed as proposed by Schönwandt, giving the approximations of the joint velocities $\dot{y}_k(t) = s_{k1}$, $k=1\dots n$, and accelerations $\ddot{y}_k(t) = s_{k2}$, $k=1\dots n$. Approximations of endeffector forces are then obtained by

$$\underline{\hat{F}}_e = \underline{J}^{T}[\,\underline{\tau} - \underline{M}\,\underline{\hat{\ddot{y}}} - \underline{h}(\underline{\hat{y}}, \underline{\hat{\dot{y}}}) - \underline{g}(\underline{\hat{y}})\,].$$

Simulations are done for a planar two link manipulator model (see e.g. [PAUL 81]) with unit lenght excluding actuator dynamics:

$$\tau_1 = (2\,m_2\,(\,1 + C_2\,) + m_1)\,\ddot{q}_1 + m_2\,(1 + C_2)\,\ddot{q}_2 - m_2 S_2 \dot{q}_2^2 - 2m_2 S_2 \dot{q}_1 \dot{q}_2$$

$$+\,g\,(m_2\,(S_1 + S_{12}) + m_1\,S_1) + j_{11} F_{ex} + j_{12} F_{ey},$$

$$\tau_2 = m_2\,(1 + C_2)\,\ddot{q}_1 + m_2\,\ddot{q}_2 + m_2 S_2 \dot{q}_1^2 + g m_2 S_{12}$$

$$+ j_{21} F_{ex} + j_{22} F_{ey}.$$

where q_i, $\dot{q}_i$, $\ddot{q}_i$, i=1,2 are the joint positions, velocities and accelerations, τ_i, i=1,2 are the joint torques, m_i, i=1,2 are the masses, j_{ij} are the elements of the Jacobian matrix, g is the gravity acceleration, F_{ex}, F_{ey} are the endeffector forces in x and y direction and $S_i = \sin(q_i)$, $C_i = \cos(q_i)$, $S_{12} = \sin(q_1 + q_2)$, $C_{12} = \cos(q_1 + q_2)$ (see fig. 5-2). Mass values are taken $m_1 = 4\ kg$, $m_2 = 2\ kg$.

The position control is a common 'computed torque'-design, there is no force control included. Different situations were simulated: 1) A step change of the forces in x- and y- directions, which can occur when the platform is accelerated; 2) Sinosoidal forces are applied, as an example for other effects; 3) contact with a rather soft environment (because of the lack of force control) is simulated using a simple spring contact model (spring constant $k_c = 1\ kN\!/m$).

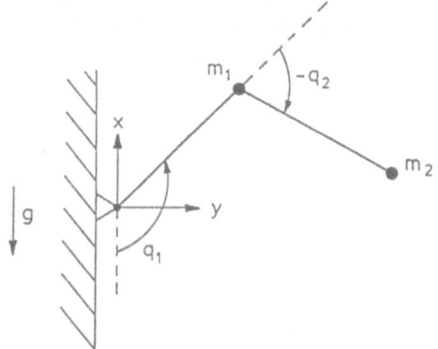

Fig. 5-2 Manipulator model

Endeffector forces were approximated using Legendre polynomials, the order of the approximation was three, the integration interval was 0.2s. In the following plots the solid lines show the real endeffector forces and the dashed lines the approximations.

Fig. 5-3 shows the result of a step change in force at time t=0.5s from 0 to 10 N in x, and from 0 to 5 N in y direction. The endeffector was position controlled at $q_1 = 135\ deg$, $q_2 = -90\ deg$. It can be seen from the simulations that after about 1s the approximated endeffector force nearly equals the true values. The final error is less than 0.1 N, what is about 1% for the x direction and 2% for the y direction. The high overshoot is due to fast changes of the output variable that can not be handled by the approximation, but for e.g. recognition of collisions the performance seems to be acceptable.

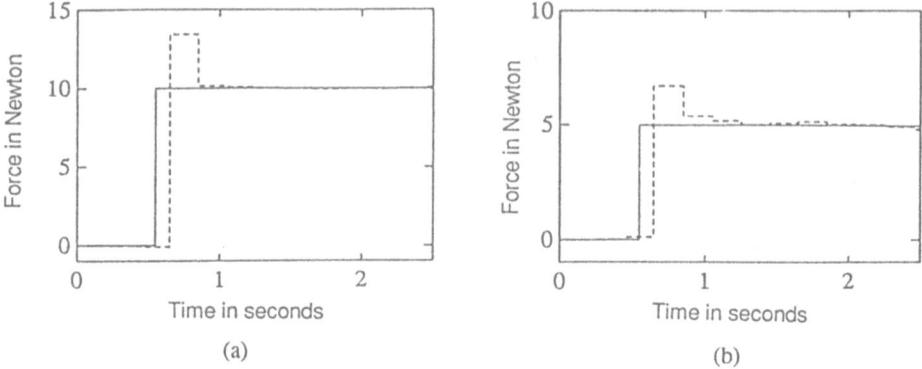

Fig. 5-3 (a)Force in x-direction, (b)Force in y-direction

Fig. 5-4 shows the result of the sinusoidal force experiment with $\omega=1$ and amplitudes of 10 N in x and 5 N in y direction. The errors between approximated and real endeffector forces are of course large during the integration, but at the time instants kT the errors are less than 5% in x and 3% in y direction.

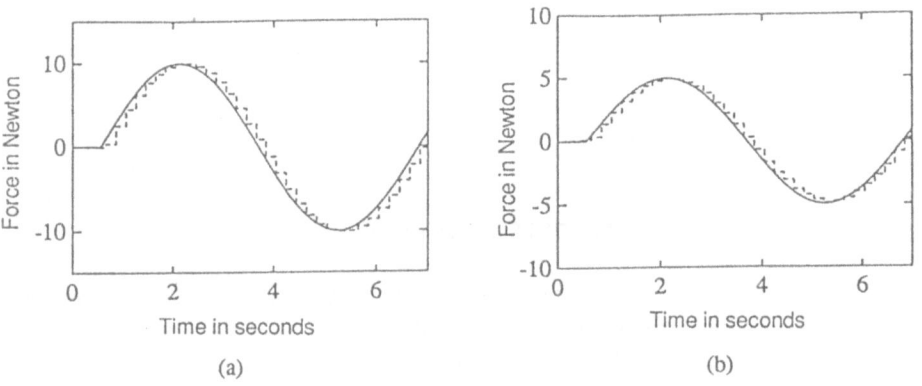

Fig. 5-4 (a)Force in x-direction, (b) Force in y-direction

In Fig. 5-5 the resulting forces for a contact experiment are shown. The robots endeffector is slowly moving in x direction and hits the soft contact surface. The movement was stopped when the endeffector force exceeded 30 N. The simulation shows that the contact forces are approximated quite good during the contact. Of course the fast dynamic changes when the movement is stopped can not be handled by the approximation leading to rather large errors. The final error is about 3%.

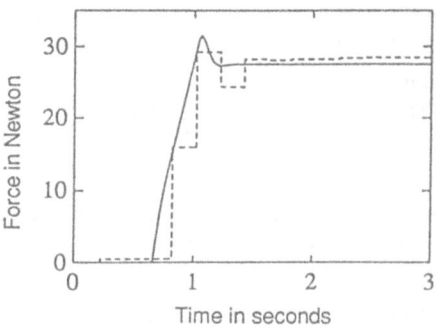

Fig. 5-5 Environmental contact

5.5 Conclusions

The simulations in 5.4 show that endeffector forces can be approximated using the technique proposed by Schönwandt. Although the approximation can be made arbitrarily good at least theoretically, it remains the problems that no convergence of the estimated endeffector forces to their true values is given by the theory, that approximated values are only available for discrete times and that the accuracy of the estimation can not be predicted. Therefore it is obvious, that the proposed technique must not be used where exact continous measurements of endeffector forces are needed. However it can be useful to determine contact situations and to be applied in contour following, especially when parts are to be inspected, where no exact knowledge of contact forces is needed, because it can be implemented quite easily, possibly in hardware, and is not very time consuming. Further investigations shall be made whether the approximated values can be used in a linearized observer model as an update or as estimates for some kind of adaptive observer.

5.6 References

[BESTLE, ZEITZ 83]
Bestle, D, ; Zeitz, M. : "Canonical Form Observer Design for Nonlinear Timeinvariant Systems"
Int. J. Control, vol. 38, no. 2, pp. 419-431, 1983

[HARTMANN, LANDGRAF 89]
Hartmann, I. ; Landgraf, Ch. : "Nichtlineare Zustandsbeobachtung und Schätzung"
Automatisierungstechnik, vol. 37, 1987

[HUANG, TZENG 89]
Huang, H. ; Tzeng, W. : "Robotic Force Control by using Estimated Contact Force"
Proc. 28th Conf. Decision and Control, Tampa, Fla., 1989

[KOU et al. 75]
Kou, S. R. et al. : "Exponential Observers for Nonlinear Dynamic Systems"
Information and Control, vol. 29, pp. 204-216, 1975

[KRENER, RESPONDEK 85]
Krener, A. J. ; Respondek, W. : "Nonlinear Observers with Linearizable Error Dynamics"
SIAM J. Control and Optimization, vol. 23, no. 2, 1985

[NAGHDY et al. 85]
Naghdy, F. et al. ; "Robot Force Sensing using Stochastic Monitoring of the Actuator Torque"
Robots and Automated Manufacture, IEE Control Engineering Series 28, 1985

[PAUL 81]
Paul, R. P. : "Robot Manipulator: Mathematics, Programming and Control"
MIT Press, Cambridge, Mass. , 1981

[SCHÖNWANDT 73]
Schönwandt, U. : "Approximations to Nonlinear Observers"
Automatica, vol. 9, pp. 349-356, 1973

[SHIMANO, ROTH 79]
Shimano, B. E. ; Roth, B. : "On Force Sensing Information and its Use in Controlling Manipulators"
Proc. 9th Intl. Symp. on Industrial Robots, Washington, D.C., pp. 119-126, 1979

[WALCOTT et al. 87]
Walcott, B. L. et al. : "Comparative Study of Nonlinear State-Observation Techniques"
Int. J. Control, vol. 45, no. 6, pp. 2109-2132, 1987

[WANG, MCCLAMROCH 89]
Wang, D. ; McClamroch, N. H. : "Position/Force Control Design for Constrained Mechanical Systems"
Proc. 28th Conf. Decision and Control, Tampa, Fla., 1989

6 Ultrasonic Modeling

L. Vietze

The first generation of mobile platforms moved along some given trajectories following magnetic fields produced by cables under the floor. This type of system would stop automatically by mechanical switches if a collision were to occur. In the next generation of mobile systems the surrounding environment around the mobile system is in addition watched with ultrasonic distance measurements. If an obstacle is detected within some security area of the mobile system, it can stop in order to avoid collision.

It can be assumed that within the next years, mobile systems will become more common outside the factories doing for example cleaning jobs. Efforts are made to alter the predefined path based on modeling the environment of the mobile system. Videocameras, lasers, radar and ultrasonic sensors can be used for this purpose. In this chapter our experiences with environment modeling using ultrasonic sensors are described.

The first section of this chapter starts with the description of the foundations of industrial airborne ultrasonic modeling. One concept of environment modeling is using an airborne phased-array-sensor which can send sharp beams in arbitrary directions. In section 6.2 some phased-array-sensor realisations suitable for modeling an industrial environment will be discussed. Two concepts of environment modeling; one based on phased-array-sensors, and the other based on multiple time of flight measurements, are discussed in section 6.3. Their experimental results, shown in section 6.4, verify these approaches.

6.1 Foundations of industrial airborne ultrasonic modeling

Ultrasonic sensors are commonly used for rangefinders in cameras, for measuring the heigth of the surface of a liquid in tanks or for remote parking meters for distances up to 10 meters. In all these applications an ultrasonic burst is emitted, reflected by objects and received, where the "time of flight" t of the first received echo is used to determine the distance of the object. Together with the known velocity of sound c the radial distance r to the object can be calculated by

$$r = \frac{1}{2} \cdot c \cdot t \quad .$$

$$(6\text{-}1)$$

For distances which are large compared to the wavelenght, spherical waves can be assumed for the ultrasonic field. For obtaining good distance accuracy of the radial distance the frequency should be as high as possible. Unfortunately, however, in air the

attenuation increases with
frequency [MASON 64]

0.7 dB/m at 23 kHz
1.8 dB/m at 40 kHz
9   dB/m at 100 kHz
33 dB/m at 200 kHz

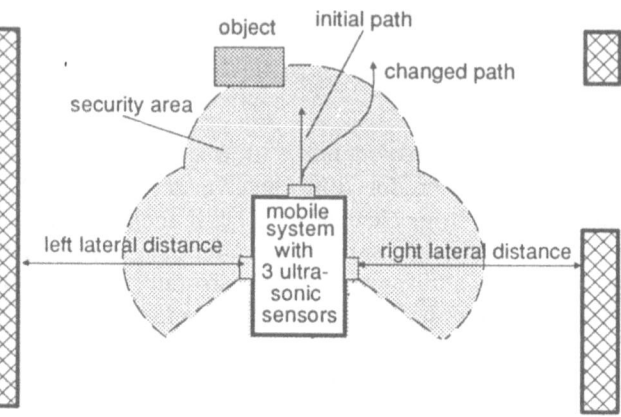

This is why, frequencies between 35 and 60 kHz are
commonly used for distan-
ces up to 10 meters [HUIS-
SON,   MOZAIRE   89],
[CIARCIAL   85]   and
[AHRENS 86] and a fre-
quency of 200 kHz for dis-

Fig. 6-1 Object detection and trajectory following

tances up to one meter [KLEINSCHMIDT, MAGORI 85], [POMEROY 85],
[SCHOENWALD 85].

The velocity of sound in air depends on air temperature, air pressure and air humidity.
The humidity of air has only a small influence on the velocity of sound; the difference
between dry air and humid air is only 0.35% of the velocity of sound at 20 degrees
Celsius. The influence of air pressure is even smaller. The velocity of sound c in air as
a function of the temperature T (with T given in Kelvin) is given by

$$c = 331.4 \cdot \sqrt{\frac{T}{273}} \cdot \frac{m}{s} \quad .$$

(6-2)

Ultrasonic systems can be used in a large field of applications for mobile robots. The
applications for mobile robots can be summarised as :

a)   Object detection:
      Object detection can be done by using simple time of flight measurements.
      Measuring the time of flight, an ultrasonic system can detect the radial distance
      of an obstacle. A security area around the mobile robot is defined by a minimal
      allowed time of flight [HINKEL, KNIERIEMEN 88]. Obstacles coming into this
      security area of the mobile robot will cause an emergency stop (fig. 6-1).

b)   Trajectory following:
      Usually the environment of the mobile robot is known in advance. In this case,
      an ultrasonic distance measuring device can be used to follow the trajectory
      according to a predefined path without any magnetic fields or optical mark-
      ings.The lateral control of the mobile system is based on two ultrasonic transdu-

cers. Additional the longitudinal position could be measured by the number of revolutions of a wheel. The first time the system has to be initialised by driving the system around to get a two dimensional description of the environment. In operation mode the system stays on a given path using two lateral ultrasonic sensors and the measured longitundial position. Because measuring the longitudinal position by the number of wheel revolutions introduces acummulation of errors the longitudinal position should be calibrated. This can be done by characteristic points in the environment. Additional ultrasonic sensors should be used in a collision advoidance system. In case the path needs be altered in order to avoid an obstacle along the predefined path of the moving robot, the lateral position of the obstacle is additionally needed (see point c).

c)   Object or environment modeling:
     The position of objects in a two or three dimensional space has to be measured for modeling the environment of the mobile robot. This information about the environment is needed for all pathplanning algorithms which are able to plan envasive actions [CHANG, QUIN 86], [LAMADRID 86]. Concepts which are used in radar- and underwater sonarsystems as well by living creatures (bats) will be transfered to airborne ultrasonic modeling systems suitable for mobile systems.

d)   Object recognition :
     Object recognition is a typical task in industrial handling. The type of object and its orientation has to be known if a robot is to pick it up [MARIOLI et. al. 88]. This can be done by comparing the received echo with several reference echos. The shape of the objects will not be recognized, only a decision is made which type (of a limited number of object types) the object under investigation may be.

Our experiences made with a phased-array-transmitter and with a multiple time of flight measurement used for environment modeling will be discussed in detail in the next sections.

6.2 Airborne ultrasonic phased-arrays

In general the sound field changes with the occurance of objects. At the position of transmitting or reflecting objects, the power of the sound field will increase. To obtain information concerning the 2D or 3D-position of objects, more than one transmitter or receiver has to be used. If the transmitters or receivers are placed at distances from each other which are large compared to the wavelength, the imaging is based on holographic models. In case the distances of the transducers are below one wavelength the transducer arrangement is called a "phased-array-sensor". A mechanically rotating sensor can be used instead of using an phased-array-sensor.The basic idea of the transmitting phase-array-sensor, also described as "antenna arrays" or "phased antenna arrays", well known from modern radar technology, is to drive an array of transmitters with different phase-shifted signals. In this way sharp formed beams can be sent in arbitrary directions without any mechanically moving parts. Phased arrays were introduced first in radarsystems to detect and track more than one of flying objects at the same time. Later on phased arrays were introduced in sonar systems and diagnostic medicine. In recent years linear (one line) airborne ultrasonic phased-array-sensors have been build for applications in robot sensing [GELLY 80], [KURODA et. al. 84], [WARNECKE 88], [LEFLEY 88] and [HUISSON 89] .

Any soundfield can be described by the time and space variable sound pressure $\underline{p}(\vec{r})$ and the acoustic velocity $\underline{v}(\vec{r})$. With the velocity of sound in air c and the transmission frequency f the wavelength λ is given by

$$\lambda = \frac{c}{f} \quad , \qquad\qquad (6\text{-}3)$$

and the angular velocity

$$\omega = 2\,\pi \cdot f \quad . \qquad\qquad (6\text{-}4)$$

If the transmitter is at the center of the coordinate system and sperical waves are assumed, the sound pressure for any position $\vec{r}$ by

$$\underline{p}(\vec{r}) = j\,\omega\,\rho\,I \cdot \frac{A(\vec{r})}{|\vec{r}|} \cdot e^{-jk|\vec{r}|} \qquad\qquad (6\text{-}5)$$

with ρ standing for air density and

$$k = \frac{2\pi}{\lambda} \quad . \qquad\qquad (6\text{-}6)$$

The factor I gives the intensity produced by the transmitting source. Positions which have the same radial distance will at the same time receive the same phase of the signal. The factor $A(\vec{r})$ takes into acount, that the soundfield of all technical sound sources depends on the transmission angle.

Literature extensively descibes the principle of the phased-array-sensor [TANCRELL 78], [SKOLNIK 80], [MONZINGO,MILLER 80], [RICHARDSON 83]. If an array of N transmitting element will be steered with different phase-shifts φ_i but the same frequency the sound pressure at any position is given using the principle of superposition by

$$p(\vec{r}) = \sum_{i=1}^{N} j \omega \rho I_i \cdot \frac{A(\vec{r}_i)}{|\vec{r}_i|} \cdot e^{-j(k|\vec{r}_i| - \varphi_i)} \quad ,$$

(6-7)

were the vector $\vec{r}_i$ stands for the vector from the ith-transmitting element to the measuring point at $\vec{r}$. Fig. 6-2 shows the measured radiation characteristic of an 8x8 airborne ultrasonic sensor (laboratory design). For estimating the quality of the phased-array-sensor the beamwidth of the mainlobe, the sending range and the power of the sidelobes has to be known. The parameters are determined by the radiation characteristics of the transducer elements, the number of transducers and the position of the transducers.

The radiation characteristic of the transducers should be unidirectional or at least have the same power in the directions of the desirable sending angles. It can be shown theoretically [KÄS 81] or by simulation, that a steerable beam in the scope of 60 degrees requires a spacing distance between the transducers of less or equal half the wavelength. A larger spacing distance limits the steerable sending angle and produces increasing sidelobes. Using a 40 kHz airborne ultrasonic sensor requires a spacing distance of d=

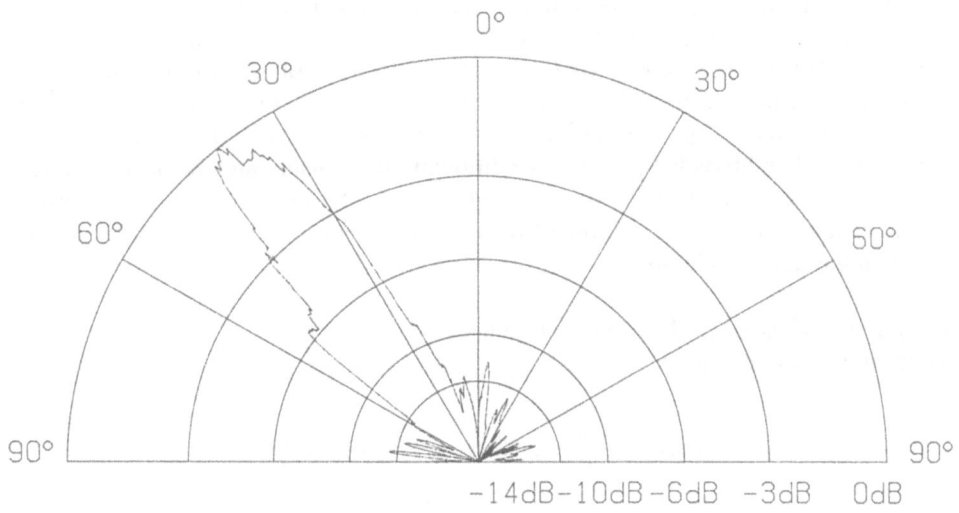

Fig. 6-2 Measured radiation characteristic of an 8x8 phased-array

4.3 millimeter. Higher frequencies
require even smaller spacing distances.

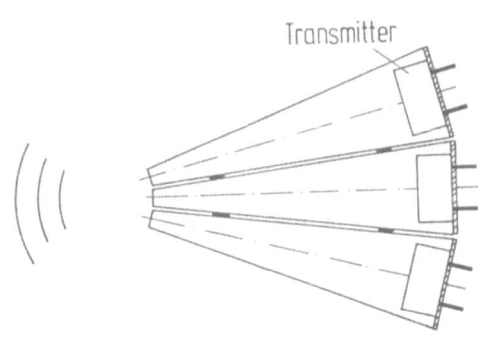

For a small number of transducer elements (less than 15 per direction)
constant spacing give good results.
Large arrays, like the ones, which are
used in tactical radar systems, perform better with stochastic spacing
distances [SANDER 82], [RICHAR-
SON, ADDISON 83], [HALFORD,
MCCULLOGH 82]. This way the en-   **Fig. 6-3** Phased-array constructed by tubes
ergy of the sidelobes can be reduced.
Because of the costs, the number of transducer elements in industrial applications will
be much smaller than in known tactical radar applications, which implies a worse
performance of the industrial systems. The description of the environment modeling
by an phased-array takes the high energy of the sidelobes into acount.

The transmitting surface of the transducers has to be less than 4x4 millimeter for a
two-dimensional phased-array. Ordinarily used ultrasonic transducers transmitting in
the frequency around 40 kHz have a diameter of 16 millimeter. Using standard
transmitters some kind of tubes for foccusing the sound beams of the transducers have
to be used. The principle of a preliminary realised labotary designed phased-array is
shown in fig. 6-3 . An example of such a sensor in an monolitic block is given by L2QZ
transducers [LÖSCHBERGER, MAGORI 87]. Transducers of this kind are constructed
in sandwich fashion from plastic wafers; additional information can be obtainted in
[KLEINSCHMIDT 85]. The sandwich fashion makes a wide range of sizes and
transmitting characteristics possible. Additionally the sensors are shock- and water-
restistant, an important feature for industrial sensors. The measured ultrasonic signals
in this chapter have been transmitted by a laboratory designed phased-array-sensor
based on the L2QZ technoligy.

The phase-shift $\varphi(x_i,y_i)$ of the transmission element at the position (x_i,y_i) depends on
the desired sending angle θ_x and θ_y by

$$\varphi(x_i,y_i) = \frac{2\pi}{\lambda} \cdot x_i \cdot \sin\theta_x + \frac{2\pi}{\lambda} \cdot y_i \cdot \sin\theta_y \quad .$$

(6-8)

For a transmission frequency of 40 kHz the wavelength λ is 8.6 mm. At some fixed
points the time difference between the transmitted and the received signal will be
measured. This time difference is equal to the time of flight of the signal and the
responce time of the transducer element. Each transducer element will give some

different phase offsets $\varphi_{off}(x_i,y_i)$ resulting from mechanical differences in the sensor design. As shown in [LANCEE 88], even small phase errors of the phased-array sensor will produce a distorted beam pattern. Phase errors are resulting from different axial positions and from different phase-characteristics of the transducer elements. For array callibration the phase-shifts of each transducer-element have to be measured. For solving this measuring task a Laser-Interferrometer can be used. The difference between the electric input signal and the motion of the transducer surface gives the phase-difference. A second possiblity would be to measrure the radiation characteristic of two neighbourhood elements. Stimulating both transducer elements with the same signal the mainlobe of the radiation characteristic of this "two-element-phased-array" should be at zero degrees. From the difference of the mainlobe from this desired output the phase difference can be calculated. In this way the whole array can be automaticaly callibrated. The transmitted phase-shifts $\varphi_s(x_i,y_i)$ is given by the sum of the calculated phase-shift $\varphi(x_i,y_i)$ and the phased offset $\varphi_{off}(x_i,y_i)$ to

$$\varphi_s(x_i,y_i) = \varphi(x_i,y_i) + \varphi_{off}(x_i,y_i) \quad . \tag{6-9}$$

The minimum allowable phase difference between two neighbourhood elements depends on the desired angular resolution. The angular resolution $\Delta\Theta_x$ and $\Delta\Theta_y$ depending on the phase-difference $\Delta\varphi_x$ and $\Delta\varphi_y$ can be calculated with equation (6-8) to

$$\Delta\Theta_x = \arcsin(\frac{\lambda \cdot \Delta\varphi_x}{2\pi \cdot x_i})$$

$$\Delta\Theta_y = \arcsin(\frac{\lambda \cdot \Delta\varphi_y}{2\pi \cdot y_i}) \tag{6-10}$$

Assuming an angular resulution of one degree, a spacing distance of 4 millimeter and a transmission frequency of 40 kHz, the minimal phase difference is 0.2 microseconds. Contrary to radar phased arrays the phase differences can be adjusted by digital TTL-logic without analog parts. The following will now discuss the generation of these phase-shifted signals.

Because of the different phase offsets of the transducers the phase-shift of each element has to be adjusted separate. For obtaining a simple phase-shift-generating hardware the usage of rectangular signals reduces the hardware effort. If a bandlimited transducer is driven by a rectangular signal, all frequencies outside the bandwith of the transducer will be erased. This can be verified by transforming the rectangular signal into the frequency domain. By definition, all frequencies outside the bandwith will heat up the transducer but will have no contribution to the transmitted energy. This means that a bandlimited transducer will transmit only a sine-shaped signal even if the transducer is driven by a rectangular signal. In many applications the long time heating up

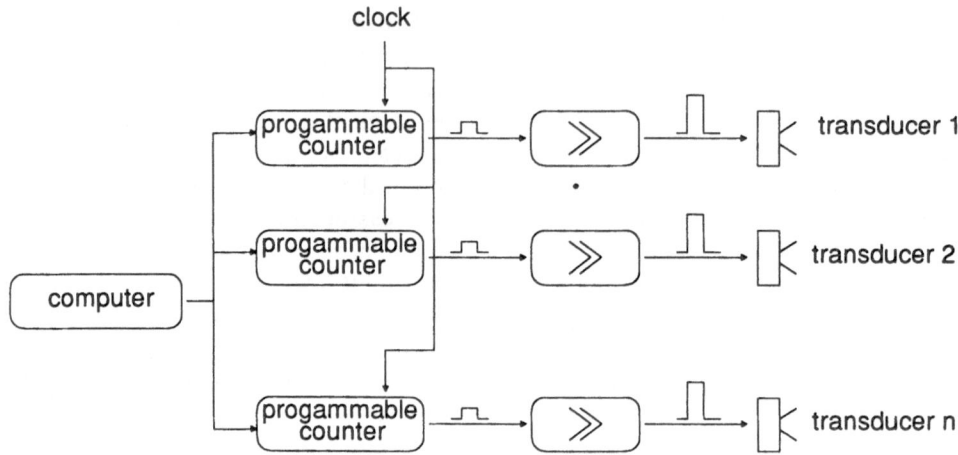

Fig. 6-4 Phase-shift-generation by programmable counters

characteristic of the transducer is the limiting factor in applying high energy pulses with a high pulse frequency. To reduce the power consumption of the transducers the edges of the rectangular signals should be smoothed by a simple passive resistance and capacitor circuit. In this way the whole phase-generation can be done by TTL-logic without analog parts.

For generating the n phase-shifted signals n programmable counters can be used. Each transducer needs it's own programmable counter which stores the phase-shift. Using 8 bit counters the minimal phase difference is

$$\Delta\varphi_x = \Delta\varphi_y = \frac{2\pi}{256} \qquad (6\text{-}11)$$

Substituting these values in equation (6-10) a minimal sending angle of $\Delta\Theta_x = \Delta\Theta_y = 0.48$ degrees can be calculated (40 kHz and 4 mm spacing distance).

Sending the signal all counters are driven by the same clock with a frequency of f=256 x 40 kHz = 10.24 Mhz. The highest bits of the counters will then change with

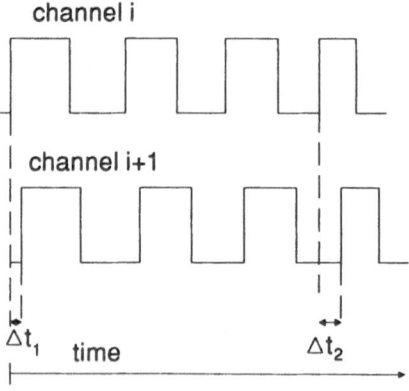

Fig.6-5 Phase-shifts for changing frequency

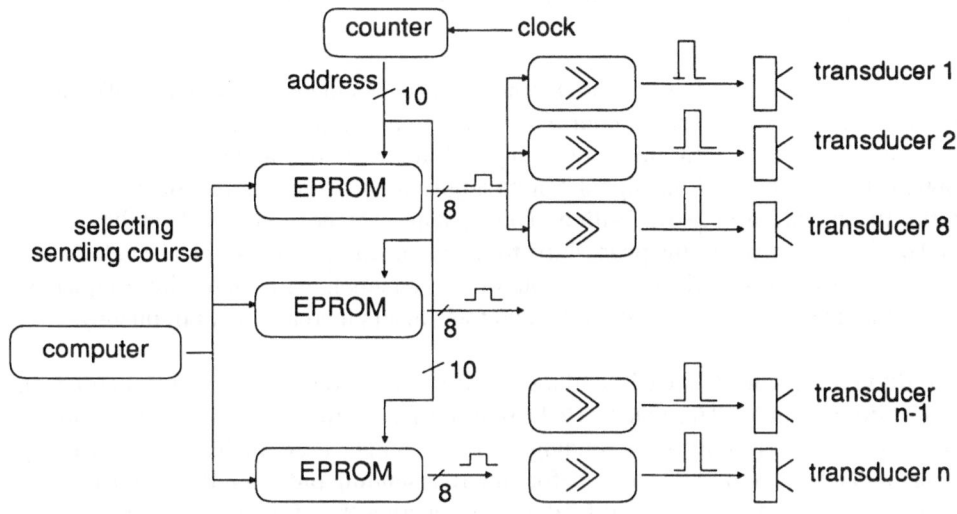

Fig. 6-6 Generation of phase-shifts by stored sending

a frequency of 40 kHz. Fig. 6-4 shows the realization of this idea where the relative phase-shifts of the counters will remain constant to each other because of the common clock. The main disadvantages of this solution are that only one transmission frequency at a time can be used and that n programmable counters have to be used. The sending direction can be changed easily, because only n bytes for the n phase-shifts have to be transmitted to generate any possible phase-shifts of the 65536 possible directions of a two-dimensional array.

In many applications of radar, medical or ultrasonic systems the performance has been increased by frequency modulated signals. Phased-array sensors impede the realization of this idea because of the changing phase-shifts at different frequencies. As it is true from equation (6-8), the change of the wavelength which corresponds to a change in frequency will result in different phase-shifted signals.

As a consequence the hardware which is able to change the transmission frequency while transmitting has to be able to change the phase-shifts continously. The transmission of a signal in a specified direction with a given signal length and frequency distribution will be defined as "sending course".

One realization which is able to change the phase-shifts continously stores the sending course of the n transducers in some memory as low and high values. To obtain the same accuracy of a steerable angle of 256 bits (=32 bytes) have to be stored for a period of a 40 kHz signal. To generate a 40 kHz signal, 128 bits will be set high, the next 128 bits will be set low. The memory, a RAM or EPROM, will be read out with a frequency of 10,24 Mhz. A counter will change the address of the readout-port to obtain the values of the sending course.

An 8kx8 RAM can be used to store 8 sending curves in the memory. The 8192 values (high or low) of each sending course are equal to a sending length of 800 µs (or 32 periods with 40 kHz). In this way changes of the frequency and phase-shifts can be obtained. The length of the signal can be adjusted by filling the rest of the memory (after the signal) with zeros. With a given spacing distance, a desired sending angle and desired frequencies, the phaseshifts for any transducer elements can be calculated. Fig. 6-5 shows schematically the output of two channels with increasing frequency. Note, that the phase-shift between the two transducers increases with frequency.

It is disadvantageous that each sending course of 32 periods requires 1 kByte for each transmitting channel. This data has to be sent from the computer device to the memory. In most applications only a few sending directions will be used. In this case the sending courses can be stored in advance, before the first sending pulse, in RAM or EPROM. Selecting the sending angle will be done by selecting the offset of the address of the memory. From this offset address, the sending values (high or low) will be read for this sending course. The schematic structure of this hardware is shown in fig. 6-6.

Receiving the echos can be done by a unidirectional receiving transducer or a receiving phased-array. The amount of parts making an ultrasonic phased-array receiver is much higher than for a phased-array transmittere because the small received analog signals of each channel have to be amplified separately by a high gain amplifier. Next, these signals have to be phase-shifted and added to the one signal.

6.3 Ultrasonic environment modeling for mobile robots

Imaging using ultrasonic signals is well known in medical applications and in sonar technology. In todays mobile systems ultrasonic sensors are used for collision avoidance. The "time of flight" of ultrasonic signals is measured to calculate the distance from the closest obstacle. If anything is within the so called security area, an emergency stop will be enforced. Mobile systems which should be able to alter the off-line determined navigation trajectory need additional information about the position of obstacles. Ultrasonic modeling should therefore be used additional to the optical information from the video inputs. In this chapter different concepts of ultrasonic modeling will be discussed. Two modeling concepts fitting to mobile systems will be described in detail.

The lateral position of objects can be obtained by scanning the environment or using holographic models. Experimental robot systems use mechanically rotating ultrasonic sensors with sharp formed beams. The time of flight of the ultrasonic signals can be determined by pulsed signals or by frequency modulated continious waves. Imaging using a phased-array-sensor will be described later on. The advantage of a phased-array-sensor is the absence of any moving parts and that the sending directions can be choosen arbitrarly without any predefined sending directions. In this way regions containing dangerous obstacles can be looked at more often than other areas.

Holographic models use the information of a couple of receivers which are placed at large distances from each other in comparison with the wavelength. Receivers placed in a linear arrangement allow a two dimensional calculation of the soundfield, receivers placed in a planar arrangement allow a three dimensional calculation.

One possibility is the usage of a continous wave monochromatic ultrasonic signal. The phase and the amplitude of the reflected signal is recorded at different receivers. With this information the soundfield of the environment will be calculated. High values of the soundfield occur at sending or reflecting objects. At all other places there will be small values of the soundfield. In this way the lateral (x-direction) and axial position (y-direction) of the objects can be calculated. Using more than one discrete frequency can improve the quality of the environment modeling.

Phase shifts of the signal occur due to small turbulences in the air and time variant velocities of sound due to small local temperature differences. This has a great effect on the calculated position of the objects. The difference of the time of flight due to temperature changes is about 0.183 % per Kelvin (see section 6.1). Assume for example, that the temperature difference of the air along two pathes to two receivers is one Kelvin. The difference of the time of flight is than for each meter 5.5 microseconds, which at 40 kHz corresponds to 22% of the wavelength.

As shown by this example the usage of the phase shift of the received signal as basic information for airborne ultrasonic signals lead to large errors. Even with the advantage of small computational costs this concept will not fit into an industrial environment. To obtain a better resolution a pulsed ultrasonic burst will be transmitted. Using holographic models [AUER 86, LÖSCHBERGER 87] the resolution can be improved compared to the discrete holography, but still small turbulences and small temperature differences will have big effects on the estimated position of the object. So that instead of using the phase "information", the time of flight of the ultrasonic burst should be the basic source of information in industrial imaging. This is true because the measuring error due to the fluctations in air procentualy is much smaller with time of flight measurements than with phase measurements. Both examined modeling concepts are based on "time of flight" measurements which will discussed in the following.

Usually a timer starts when the transmitting of the signal begins. If the received signal is above a given intensitity, the timer stops and the time of flight is given. To minimize the effects of noise usually bandpass filtering and averaging over a couple of measurements is done. Better results can be obtained by using additional information about the reflection process. Reflection of an ultrasonic burst on a large flat plate is shown in fig. 6-7 . This burst was generated by 10 rectangular pulses with a frequency of 40 kHz. As known, diffuse reflection on smooth surfaces as in an industrial environment can be neglected. In this way the received signal will be assumed as an ensemble of echos from flat reflecting plates. The received echo $s_n(t)$ can be written approximately as a superposition of p normalized echos $s_{ref}(t)$ with different time of flights and different power coefficents a_i and the noise $n(t)$ as

$$s_n(t) = n(t) + \sum_{i=0}^{p} a_i \cdot s_{ref}(t-t_i)$$

(6-12)

To reduce the effects of noise, the signal $s_n(t)$ should be bandpass filtered as a first processing step. The power coefficient a_i depends on the reflection coeffient a_{ri} of the reflecting obstacle, the radiation characteristic into the obstacle direction a_{ci} and the propagation coefficient a_{si} due to the spherical propagation and the frequency dependent attenuation in air. Multiplying these coefficients the power coefficient a_i can be calculated as

$$a_i = a_{ri} \cdot a_{ci} \cdot a_{si}$$

(6-13)

The reflection coefficient a_{ri} depends on the size of the object, the structure of its surface and the position. If the measuring scene is nearly stationary and the mobile system moves slowly the reflection coefficient a_{ri} and the propagation coefficient a_{si} will not change. This statement should be kept in mind for the later described signal

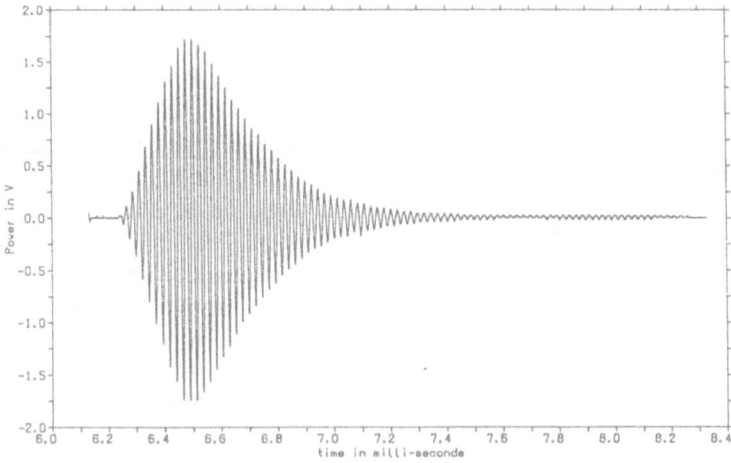

Fig. 6-7 Reference echo

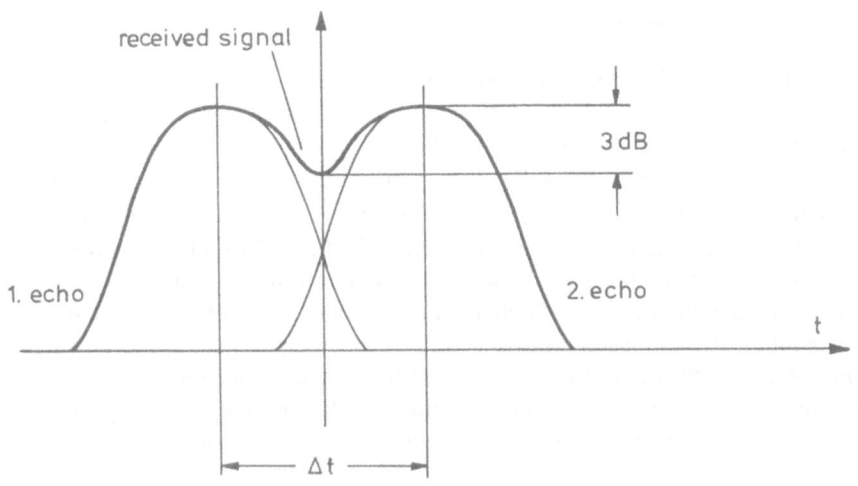

Fig. 6-8 Resolution in ultrasonic imaging

processing. Equation (6-12) is the basis for filtering the received signal. It can be shown [VAN TREES 68], that an optimal solution of detecting a known signal $s_{ref}(v \cdot T)$ of lenght N in a disturbed signal $s_n(v \cdot T)$ which is added by white noise is given using crosscorrelation function

$$R_{xy}(v \cdot T) = \frac{1}{N} \sum_{n=0}^{N} s_n([n+v] \cdot T) \cdot s_{ref}(n \cdot T) \quad .$$

(6-14)

The times of flight from transmitter to the objects and back to the receiver, are given by the maxima of $|R_{xy}(v \cdot T)|$:

$$t_i = v_i \cdot T \quad .$$

(6-15)

In order to separate two echos the intensity of the first echo has to be dropped below a specific value. Usually this value is chosen to be 3 dB below the intensity of the first maximum (fig. 6-8). It can be shown, that an increasing bandwith of the transducer will speed up the building-up time of the transmitter signal. This improves the separation of two close objects [LASSAHN, BAKER 82].

In the following two sections two concepts of modeling the environment based on time of flight measurements using the phased-array-sensor and a couple of sensors at different positions are discussed.

6.3.1 Modeling using an phased-array-sensor

To obtain a model of the environment with the use of a phased-array-sensor the sensor has to scan around. The following ideas will be demonstrated in an empty test room containing only one post. For each sending direction j the receiver will receive the reflected signal $s_{n,j}$ (t). In fig. 6-9 the intensity of the crosscorrelation function given by equation (6-14) $| R_{xy} (v \cdot T , j) |$ obtained with the signal of an 8x8 phased-array is drawn as a function of the sending angle and the time of flight. At the position of the highest maximum the position of the post in polarcoordinates can be seen easily by a human. All the other maxima which having the same "times of flight" are produced by the sidelobes of the phased-array-sensor. Because of these sidelobes no information about the lateral position of the objects can be gained by one measurement with only one sending direction. An accurate modeling is only possible after scanning the whole scene under the assumption that the scene does not change while scanning.

From a general point of view the "true" maxima has to be determined for a two dimensional curve, without being affected by disturbances or the effects of sidelobes. As seen in the figure more than one maximum will occur because of the effects of the sidelobes which are about -10dB for an 8x8 phased-array. Cutting everything below 10 dB of the largest maxima is dangerous, because smaller echos from other objects will get lost.

The maxima produced by the sidelobes of the phased-array-sensor are always at the same radial distance as the objects. To suppress the effect of the sidelobes, the transmitting characteristic of the phased-array-sensor has to be known. Usually only a discrete number of angles are used while scanning. To record the transmitting characteristic of the phased-array, a reflecting object, e.g. a post, is moved in discrete steps around the phased-array-sensor with the same axial distance. The reflected signal $s(v \cdot T, \theta_x, \varphi_i)$ for each position (r,φ_i) of the post (in polarcoordinats) and each sending angle θ_x will be recorded. The maximum of the crosscorelation function

$$R_{\theta_x,\varphi_i} (v \cdot T) = \frac{1}{N} \sum_{n=0}^{N} s ([n + v] \cdot T , \theta_x, \varphi_i) \cdot s_{ref} (n \cdot T)$$

$$(6\text{-}16)$$

gives the reflecting characteristic $\hat{R}_r(\theta_x, \varphi_i)$ as a two dimensional array depending on the sending angle and the angular position of the object. To simplify the recording of the reflecting characteristic the following measurement arrangement can be used. The phased-array is mounted on a stepper motor. Instead of moving the post at the same axial distance around the phased-array, the phased-array is being swung round the vertical axis. In this way the reflecting characteristic can automatically be obtained for each sending angle and each of the angular directions of the reflector.

To suppress the sidelobes, the intensity of the crosscorelation function $| R_{xy} (v \cdot T, j) |$ will be correlated for each of the j transmitting directions θ_x with the reflecting characteristic $\hat{R}_r(\theta_x, \varphi_i)$ to

$$S_r (v \cdot T, \theta_x) = \frac{1}{j} \sum_{n=0}^{j} | R_{xy} (v \cdot T, n) | \cdot \hat{R}_r (\theta_{x}, n) \quad .$$

(6-17)

To reduce the calculational burden this function will only be calculated for the maxima of the the crosscorrelation function from equation (6-14).

In each sending directions all maxima of equation (6-17) and their positions are stored. In the next step all maxima which are produced by the same object are attached. Because of fluctations the estimated times of flight will differ slightly. Ordering has to be done by selecting the maxima within the before defined "3dB-range". For each "time of flight" on which a maximum occurs a curve of the estimated maxima will be given. At this curve all maxima which are below a special level will be erased. As a result only the maxima of the objects will remain. With this idea objects with a different time of flight but very different reflecting intensity will be detected.

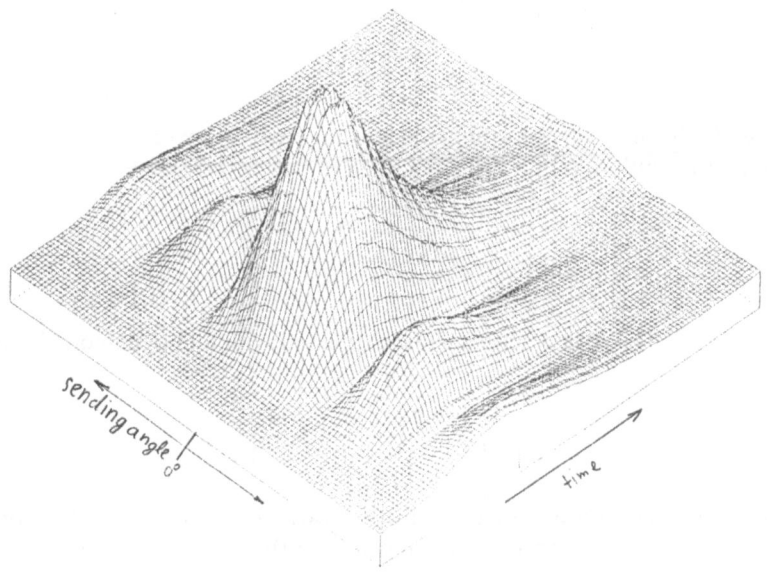

fig. 6-9 Crosscorelation function of the scanned environment

6.3.2 Modeling by different time of flight measurements

If an obstacle is getting closer to the ultrasonic sensor an echo will be received within a defined time-intervall. Using the same transducer as transmitter and receiver the possible place of the object is on a circle. The radius of the circle can be determined by the measured time of flight. If different transmitters and receivers are used, the place of the obstacle is given by an ellipse. Using two receivers the place of the object can be calculated with the different time of flights. This idea, which is used by bats and humans [KAY 79], [ESCUDIE 79], [VIETZE, HARTMANN 90], [WIDJAJA 91] is very robust against fluctuations of the measured times of flight.

Assume the transmitter is at the position X_0 (x_0 , y_0) and the receiver i at the position X_i (x_i , y_i). The first object is at the unknown position P (x_p , y_p).With the velocity of sound c the distance from the transmitter to the object and back to the receiver is given by

$$\overline{X_0 P} + \overline{P X_i} = c \cdot t_i \quad . \tag{6-18}$$

This equation (6-18) describes an ellipse with the center at $(x - \dfrac{x_i}{2} , y_0)$

$$\frac{(x - \dfrac{x_i}{2})^2}{a_i^{\,2}} + \frac{y^2}{b_i^{\,2}} = 1 \quad . \tag{6-19}$$

In fig. 6-10 the measuring arrangement using two receivers is shown. From the definition of the ellipse (2a = l) the parameter

$$a_i^{\,2} = \frac{t_i^{\,2} \cdot c^2}{4} \tag{6-20}$$

can be deduced. The parameter b_i can be expressed with thePpythagoras's equation to

$$b_i^{\,2} = \frac{t_i^{\,2} \cdot c^2}{4} - \frac{x_i^{\,2}}{4} \tag{6-21}$$

Each additional receiver gives a new ellipse on which the position of the object has to be. With two receivers placed at coordinates $(x_1,0)$ and $(x_2,0)$ the measured time of flight t_1 and t_2 can be used to obtain the parameters a_1, b_1, a_2 and b_2 . At the intersection of the two ellipses the x- and y-coordinate of the two ellipses has to have the same

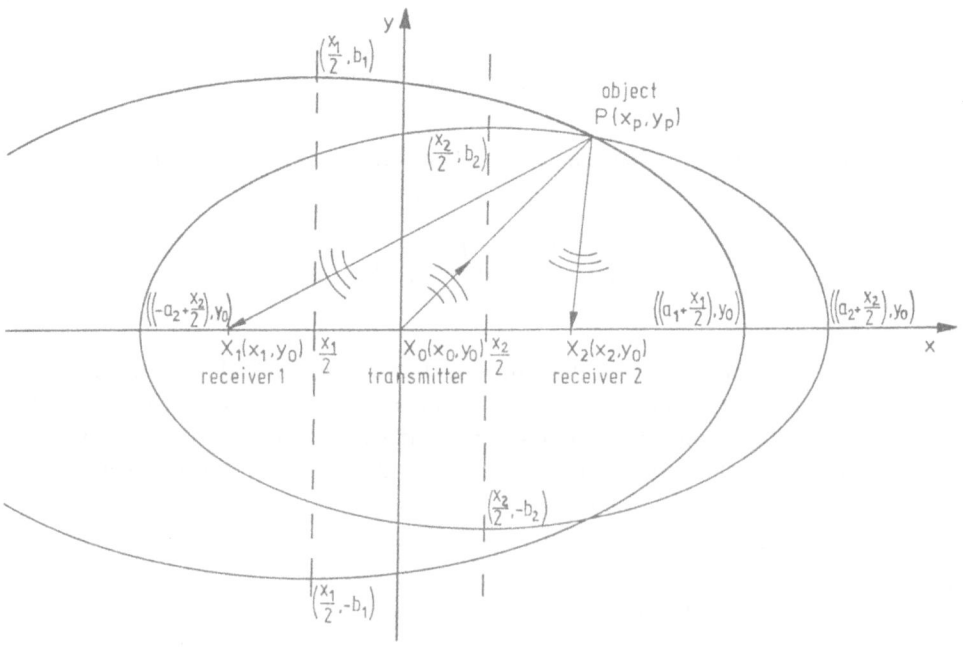

Fig. 6-10 Time of flight measuring arrangment

value. The x_{si} coordinate of the Intersection with the receiver at the places x_1 and x_2 has to fit the equation

$$\left\{ 1 - \frac{(x_{si} - \frac{x_1}{2})^2}{a_1^{\ 2}} \right\} \cdot b_1^{\ 2} = \left\{ 1 - \frac{(x_{si} - \frac{x_2}{2})^2}{a_2^{\ 2}} \right\} \cdot b_2^{\ 2} \tag{6-22}$$

which can be obtained by equation (6-19). The result is an quadratic equation

$$(a_1^{\ 2} b_2^{\ 2} - a_2^{\ 2} b_1^{\ 2}) \cdot x_{si}^{\ 2} + (x_1 a_2^{\ 2} b_1^{\ 2} - x_2 a_1^{\ 2} b_2^{\ 2}) \cdot x_{si} +$$

$$(a_1^{\ 2} a_2^{\ 2} b_1^{\ 2} - a_2^{\ 2} a_1^{\ 2} b_2^{\ 2}) +$$

$$(\frac{x_2^{\ 2}}{4} a_1^{\ 2} b_2^{\ 2} - \frac{x_1^{\ 2}}{4} a_2^{\ 2} b_1^{\ 2}) = 0 \quad . \tag{6-23}$$

Solving this quadratic equation two value x_{si} will be obtained, even there is only one true value. The estimated position x_{si} has to fit the geometrical inequality

$$- a_1 + \frac{x_1}{2} \leq x_{si} \leq a_2 + \frac{x_2}{2}$$

(6-24)

what is given only for one value. The estimated position y_{si} is given by the root of

$$y_{si}^2 = \left\{ 1 - \frac{(x_{si} - \frac{x_1}{2})^2}{a_1^2} \right\} \cdot b_1^2 \quad ,$$

(6-25)

Because the objects are in front of the ultrasonic transducers, only the positiv value of y_{si} is valid. Increasing the number of receivers improves the accuracy of the position measurement. Using for example four receivers the four ellipses will result in six intersections. The estimated position of the object $P_s(x_s, y_s)$ can be obtained by minimising the criteria least-square

$$Q = \min \left\{ \sum_{i=1}^{6} \sqrt{(x_s - x_{si})^2 + (y_s - y_{si})^2} \right\} \quad .$$

(6-26)

6.4 Experimental results

Modeling the environment using the phased-array-sensor and the times of flight. has been implemented and tested under a various number of testing surroundings. To produce comparable results, round posts are used. The position of more complex objects, e.g. humans, can also be calculated. But it is difficult to give the accuracy of the position for such complex surfaces.

6.4.1 Results with the phased-array

The following ideas will be demonstrated in an empty test room containing three posts at different places. Two of the posts have the same axial distance to the sensor but different lateral positions, the third post is behind them. In fig. 6-11 the crosscorelation function of the scanned environment is shown. The limits of accuracy of modeling the environment by a phased-array-sensor is given by the accuracy of the time of flight measurements and the difference of the reflected intensity while scanning. Using a couple of sending directions the accuracy of the the time-of-flight measurements can be improved by averaging the different time of flight measurements. In order to

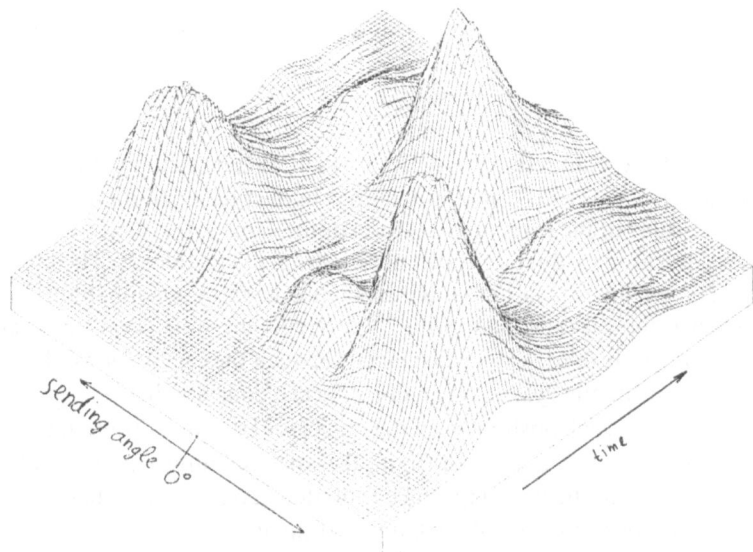

Fig. 6-11 Crosscorrelation function of the scanned environment

compensate for the temperature dependent velocity of sound, the temperature should be measured to calculate the correct velocity of sound. In this case the accuracy of the axial distance is about ± half the wavelength which at 40 kHz is equal to ± 4.3 mm. Due to fluctuations of the reflected intensity of the signal, the maxima of the crosscorrelation function is not always the middle of the object. The accuracy of the angular position of a post is for our 8x8 phased-array ± two degrees.

6.4.2 Results with the time of flight measurements

As discussed in section 6.4.1 the estimated time of flights differ slightly around the average value if the temperature is known. In the following a theorectical approach of the accuracy of the position will be given. The measuring arrangement is shown in fig. 6-12. The transmitter is in the center of the coordinate system $X(x_0, y_0)$, the two recievers are at the places $X_1(x_1, y_0)$ and $X_2(x_2, y_0)$. After transmitting one burst two different time of flights t_1 and t_2 for the two receivers can be calculated. With the known velocity of sound c the distance l_{t1} and l_{t2} from the transmitter to the object and back to the receivers can be calculated by

$$l_{ti} = \overline{X_0 P} + \overline{P X_i} = c \cdot t_i \quad . \tag{6-27}$$

The true distances l_1 and l_2 are given with the measurement errors Δl_1 and Δl_1 by

$$l_i = l_{ti} + \Delta l_i \quad . \tag{6-28}$$

The accuracy of the y-direction is given for small distances x_p by the accuracy of the time-of-flight measurements. Small fluctuations in the time of flight measurements will result in fluctuations of the estimated position of the same size. However the fluctuations of the lateral position (x-direction) can not be so easily calculated. The basic information for obtaining the x-position is the difference of the times of flight. To illustrate this, a measuring arrangement with two receivers at $X_1(x_1 = -350mm, y_0)$ and $X_2(x_2 = 350mm, y_0)$ is used. Larger distances between the receivers would be nice, but will not fit on our mobile system. If an object is at a constant distance y_p and is moved paralell to the x-axis each position x_p of the object will result in different times of flight. The difference between the times of flight, the essential information for obtaining the lateral position, versus the lateral distance is plotted in fig. 6-13. To make the plot easier to read, the times of flight are calculated in millimeters.

The accuracy of the time of flight measurements can be easily derived from experiments for a number of various objects and distances. With this information, the fluctuations around the true lateral position of the object can be estimated. An increasing gradient of the curve in the estimated place $\hat{P}(x_s, y_s)$ of the object result in a better lateral accuracy. With this idea the area in which the object has to be located is given.

To verify this idea, time of flight measurements have been made with different reflecting objects. The echo of the better reflecting object (a 50 mm wide copperpost) was six times stronger than the echo of the second object (a 20 mm round post). Experiments shows, that the accuracy of the time of flight measurements depend strongly on the power of the reflected signal.

The signal to noise ratio depends for a given transmitting energy, a given noise-energy and a constant distance only on the reflecting coefficients of the objects. Using the good reflector (the copperpost), the accuracy of the time of flight measurements are at a distance of 2 meters ± 4 millimeter (equal ± half the wavelength).The result for the weak reflector in the same distance gave fluctuations of ± 15 millimeter (equal ± two wavelength). Experiments show, that with a known velocity of sound the fluctuations are sym-

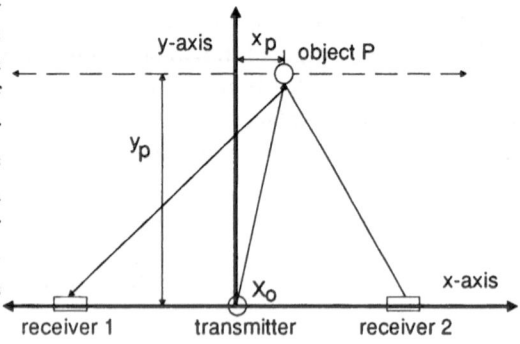

Fig. 6-12 Measuring arrangement

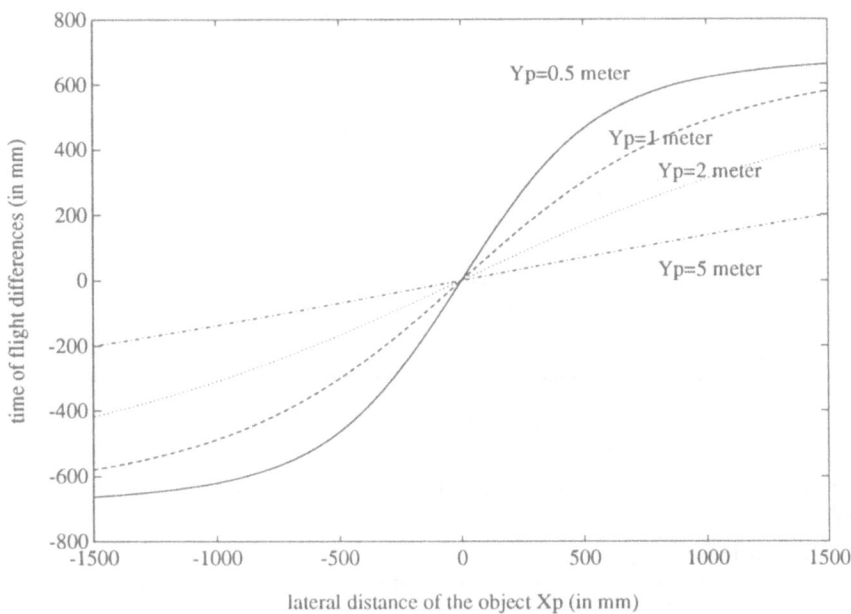

Fig. 6-13 Time of flight differences for different object distances

metriacal around the true value. Using an average of a couple of measurements, the accuracy can be improved. Experiments verify this approach.

6.5 References

[AHRENS,LANGEN 86]
Ahrens, U.; Langen, A. : "Ultraschallscanner- Ein neues System zur Roboterführung"
VDI-Berichte Nr.518, VDI-Verlag 1986 ; pp. 337-350

[AUER 86]
Auer, B.: "Bildkonstruktion kleiner Körper durch digitale Verarbeitung reflektierter Ultraschallsignale"
P.D. Thesis TU Berlin 1986

[BERKTAY, ORHAN 85]
Berktay,H.; Orhan: "Resolution in Sonar Systems-A Review"
Proceedings of the 14.th. International Symposium on Acoustical Imaging; 22.-25. April 1985; Plenum Press, New York 1985

[CHANG, QUIN 86]
Chang, T.S.; Quin, K. : "An obstacle avoidance algorithm for an autonomus land vehicle"
Proceedings of SPIE 727 of Mobile robots; 30-31.Oct. 1986; pp. 117-123

[CIARCIA 80]
Ciarcia, S.: "Home in in the range! An Ultrasonic Ranging System"
Byte Publication Inc. Nov 1980

[ESCUDIE 79]
Escudie,B.: "Signal Processing and Design Related to Bat Sonar Systems"
Animal Sonar Systems (Ed. Busnel, R.G. and Fisch, J.F.) pp. 715-729; Plenum Press New York 1979

[GALEY, HSIA 80]
Galey, B.; HSIA, P. : "A Survey of Robot Sensor Technology"
12 th. Annual Southeastern Symposium on System Theory; May 19 and 20 1980; pp. 90-93

[GELLY 81]
Gelly, J.F.: "Properties for a 2 D Multiplexed Array for Acoustical Imaging"
IEEE Ultrasonics Symposium 1981; pp. 685-689

[HALFORD, MCCULLOGH 82]
Halford, G.I., McCullogh, W.J. : "Minimisation of sidelobes from a planar array of Uniform
Elements"
International Conference of Radar-82; 1982 IEEE; pp. 360-364

[HAYWARD, GORFU 89]
Hayward, G.; Gorfu : "Low power ultrasonic imaging using digital correlation"
Ultrasonics 1989 Vol.27; september 1989; pp. 288-296

[HINKEL, KNIERIEMEN 88]
Hinkel, R. ; Knieriemen, Th. : "Environment Perceptron with a Laser Radar in a Fast Moving
Robot"
Preprint of the IFAC-Symposium Robot Control 1988; Oct. 5.-7. 1988; pp. 68.1-68.7

[HUISSON, MOZIAR 89]
Huisson, J.P.; Moziar, D.M.: "Curved ultrasonic array transducer for AGV applications"
Ultrasonics 1989 Vol. 27 July 1989   pp. 221-225

[KAY 79]
Kay, L.: "Air Sonars with Acoustical Display of Spatial Information"
Animal Sonar Systems (Ed. Busnel, R.G. and Fisch, J.F.) pp. 715-729; Plenum Press New York 1979

[KAY 85]
Kay, L.: "Airborne ultrasonic imaging of a robot work space"
Sensor Review (GB) Jan 1985, vol.5, no.1, pp. 8-12

[KÄS 81]
Käs, G.: "Radartechnik]
Expert Verlag Grafenau/Württ 1981

[KLEINSCHMIDT, MAGORI 85]
Kleinschmidt, P.; Magori, V.: "Ultrasonic Robotic-Sensors for Exact Short Range Distance
Measurement and Object Identification"
IEEE 1985 Ultrasonic Symposium; Oct 16-18, 1985 San Francisco,CA

[KURODA et. al. 84]
Kuroda, S.; Jitsumori, A.; Inari, T. : "Ultrasonic Imaging System for Robots using an Electronic
scanning"
Robotica (GB) Jan. 1984; Vol. 2; pp. 47-53

[LAKIN 80]
Lakin, K.M.; Sheppard, W.R.; Tam, K.: "Acoustic Imaging with Two Dimensional Arrays"
IEEE 1980 Ultrasonic Symposium pp. 738-741

[LAMADRID 86]
Lamadrid, J.G. de : "Avoidance System for Moving Obstacles"
Proceedings of SPIE 727 of Mobile robots; 30-31.Oct. 1986; pp. 304-311

[LASSAHN, BAKER 82]
Lassahn, G.D.; Baker, A.G. : "Errors in Cross-Correlation Peak Location"
Journal of Dynaics Systems, Measurement, and Control; Vol. 104; June 1982; pp. 194-199

[LEE, FURGASON 85]
Lee, B.B; Furgason, E.S.: "The use of Correlation Systems for Real-Time Ultrasonic Imaging"
Proceedings of the 14 th. International Symposium on acoustical Imaging 22.-25 April 1985

[LEFLEY 88]
Lefley, P.W., Blanchfield, P.; Brycer,G.W.: "Heuristically guided phased array for robot controll"
ISATA 19 th. International Symposium on Automatic Technology and Automation 1988; Monte
Carlo, Monaco 24.-28.Oct. 1988; pp. 357-369

[LÖSCHBERGER 87]
Löschberger, J.R.: "Ultraschall-Sensor-System zur Bestimmung axialer und lateraler Strukturen mit
Hilfe bewegter Wandler zum Einsatz in der industriellen Auftomatisation"
P. D. Thesis Universität der Bundeswehr München 1987

[LÖSCHBERGER,MAGORI 8]7
Löschberger, J.; Magori, V.: "Ultrasonic Robotic Sensor with Lateral Resolution"
Ultrasonic Symposium 1987 14.10-16.10.87 Denver, CO

[MARIOLI et. al. 88]
Marioli, D.; Sardini, E.; Taroni,A.: "Shape determination and robot arm control positioning by
means of ultrasonics"
Proceedings of the 7. th. International Conference on Robot Vision and Sensory Controls; Feb. 2.-4.
1988; Zürich; pp. 171-182

[MASON 64]
Mason, W.P.: "Physical Acoustics"
Academic Press 1964

[MONZINGO,MILLER 80]
Monzingo, R.A. ; Miller, T.: "Introduction to Adaptive Arrays"
John Wiley & Sons New York, 1980

[POMEROY et. al. 85]
Pomeroy, S.C.; Dixon, H.J.; Wybrow, M.D.; Knight, J.A.G.: "Ultrasonic distance measuring and
imaging systems for industrial robots"
Proc. of the 5 th. International Conference on Robot Vision and Sensory Controls, 29-31.Oct 1985;
Amsterdam; pp. 239-249

[RETIEG,HAKKESTEEGT 88]
Retieg, P.P.L.; Hakkesteegt, H.C.: "A low-cost sonar system for object identification"
Proc. of the 18 th. Symposium on Industrial Robots, pp. 201-210; April 1988

[RICHARDSON,ADDISION 83]
Richardson, J.M. ; Addision, R.C.: "Acoustical Arrays with nonuniform Element Spacing and / or
elevation"
IEEE Ultrasonic Symposium 1983 ; pp. 1043-1040

[SANDER 82]
Sander, W.: "Beam Forming with phased array antennas"
Radar-82, IEE 1982 London; pp. 403-407

[SASAKI, TAKANO 88]
Sasaki, K.; Takano,M.: "Ultrasonic range sensor assists 6-DOF manipulator by locating objects in
3-dimensional space"
Proceeding of the International Symposium on Industrial Robots, April 1988; pp. 211-220

[SCHOENWALD 85]
Schoenwald, J.S.: "Strategies for Robotic Sensing using Acoustics"
IEEE 1985 Ultrasonic Symposium; pp. 472-482

[SKOLNIK 80]
Skolnik, M.I.: "Introduction to Radar Systems"
McGraw-Hill Book Company New York 1980

[TANCRELL 78]
Tancrell, R.H.: "New Field Transcient Acoustic Beam Forming with Arrays"
Proceedings of the IEEE Ultrasonic Symposium 1978; pp. 339-343

[VAN TREES 68]
Van Trees, H.L.: "Detection, Estimation and Modulation Theory"
John Wiley and Sons, New York 1968

[VIETZE, HARTMANN 89]
Vietze, L. ; Hartmann, I.: "An Ultrasonic Phased-array-Sensor for Robot Environment Modeling
and fast Detection of Collision Possibility"
Proceedings of the INCOM'89 ; 26.-29.9.89 in Madrid; pp. 373-378

[VIETZE, HARTMANN 90]
Vietze, L.; Hartmann, I.: "Ein Transputersystem als Prozessrechner zur Modellierung der
Umgebung mit Ultraschall"
2. Transputer-Anwender-Treffen 17.-18.9.90; Informatika-Fachberichte; Springer Verlag; 1990

[WARNECKE,LANGEN 88]
Warnecke,H.J.; Langen,A.: "New ultrasonic sensors for robit applications based on beam forming"
Proceedings of the International Conference on Robot Vision and Sensory Control; Feb. 1988; pp.
149-160

[WIDJAJA 91]
Widjaja, H.: "Bestimmung von Objektkoordinaten mit Ultraschallsensoren auf einem Transputer-
system"
Studienarbeit : Institut für Regelungstechnik und Systemdynamik; TU Berlin 1991

7 Lane Recognition and Following

J. Moebius

7.1 Introduction

One of the most important parts of autonomous vehicle-control are the determination of environment information. For different mission tasks there is special information needed, but in any case information about a possible path or - in structured environments - the admissable lane is necessary.

This chapter presents two pattern recognition algorithms to detect the roadway in the drivers display. Both are using picture processing methods. No subject of this chapter are other topics in information extraction such as obstacle detection or special mission tasks.

The sensor for both algorithms presented is a black and white camera. This is only one of the possible sensor principles. The good human performance in car-driving is the basis for this reasonable assumption. The human driver gets most of the information needed with his eyes. In the physical sense this is an optical stereo system with a spectral sensitivity of 400 to 700nm. If we use only one camera, we loose the depth information. This is possible for roadway detection. For obstacle detection the depth information can be essential. At night, the human being cannot distinguish colors, nevertheless he is able to control the vehicle. So it seems possible to use only one black and white camera, which substantially reduce the amount of data to be processed.

Autonomous vehicles operate in different environments. There is a wide range of difficulties to detect the roadway in these situations. Fig.7-1 and fig.7-2 clarify these difficulties by showing two different driver displays of public streets.
The algorithms presented work successfully for situations up to the difficulty of motorways or highways. In these cases the road edges are mostly visible. In contrast in urban situations the edges are often hidden by other cars or houses. For the former situations one can solve the problem in real-time today. This means that it is possible to realize a sample rate which fits the vehicle dynamics. The task of roadway detection can be described as segmention. The goal is to segment the picture in the regions "road" and "non-road". Fig.7-3 shows the required segmentation for one display. There are no difficulties for a human being to solve the described tasks. The human being has a very good performance in optical picture interpretation. In contrast it is a challenge for electronic picture processing systems.

Fig. 7-1 Highway

Fig. 7-2 Downtown

It is possible to group the large number of possible ways to detect the roadway: Direct algorithms use features of the edge, e.g. the grey value gradient. To take advantage of a-priori-information edge-following algorithms can be used. Indirect algorithms segment the picture by first splitting it into small uniform parts und second by merge these parts to groups (split and merge). The desired edge line is the dividing line between both regions.

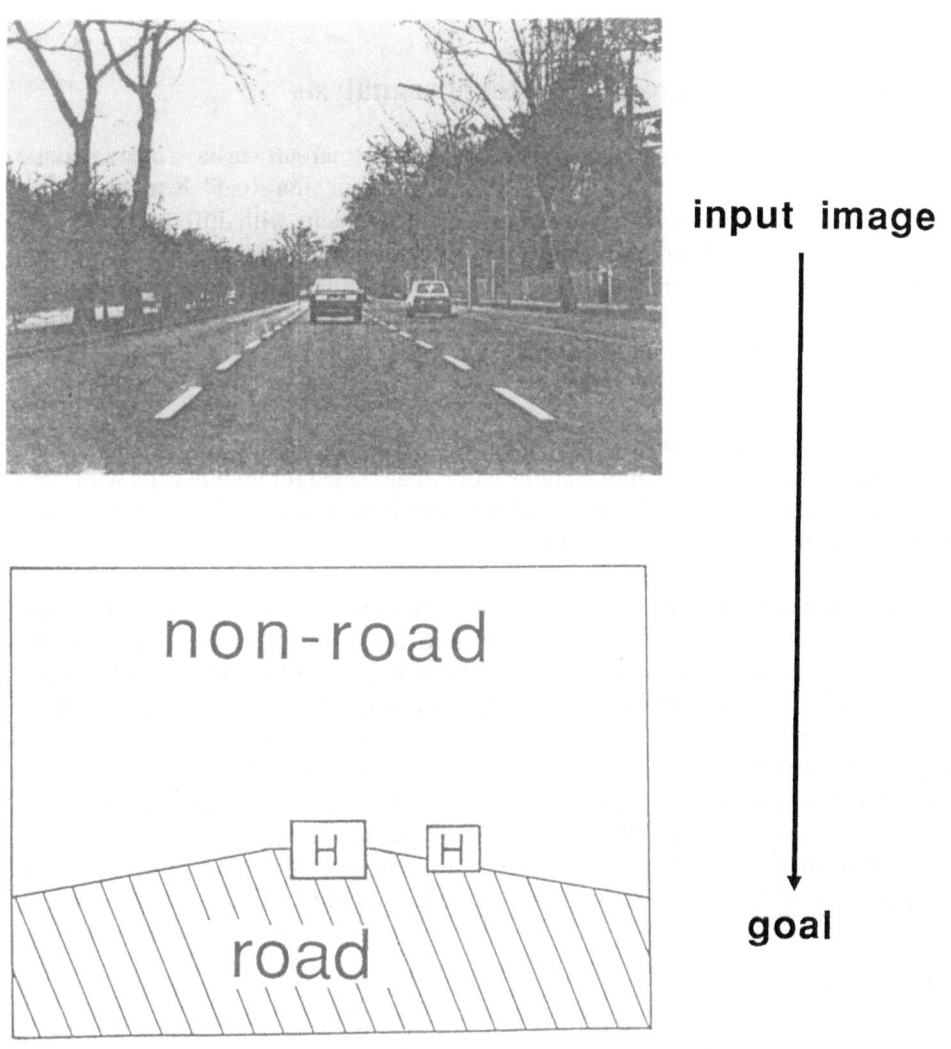

Fig. 7-3 Display, Segmentation

In the following sections two algorithms are presented, one of the direct and one of the indirect way. Edge detection by texture analysis belongs to the indirect group, the PDAF (probabilistic data association filter) to the direct group. This is an edge following algorithm, which has a big similarity to the well known Kalman filter.

7.2 Indirect edge detection by texture analysis

The starting point for this approach is the idea that each road surface has a characteristic texture, which is different from the texture of the region "non-road", here also called background. The background will be composed of segments with different texture. In contrast, the street will have a homogeneous texture for long distances. Two points have to be taken into consideration: First, the texture will change unsteadily at points of changing road surface. Second, the projection of the scene into the driver display will shorten wide distance texture elements. It is no problem to adapt the algorithm to these situations.

Texture is a property of a neighborhood of pixels, a texture feature for one single pixel cannot be defined. So the texture features will be calculated for texture cells with sizes from 8x8 to 30x30 pixels. A higher number of pixels per texture cell results in a better accuracy but in larger computation times.

The algorithm consists of the steps feature extraction and classification, as known from every pattern classification problem. The classification step needs information of the feature distribution in the distinct classes. This information has to be calculated in a learning or initialization phase. The texture of the background changes frequently in most of the scenes, so it is impossible to estimate characteristic features for the class background. This fact influences the design of the classificator. The texture of the street changes only at points of changing road surface. With the assumption that the vehicle's position is on the roadway in the initialization phase, road texture is found in a special area confidently (see figure 7-3). In this area the features can be estimated in each sampling period. So it is possible to adapt the algorithm to a changing road surface.

7.2.1 Feature extraction

Different features are known to describe textures. They can be classified in frequency-domain and spatial-features. Here we use spatial features, published first in [HARALICK et. al. 73]. This approach starts with the quadratic spatial-dependence-matrix (SDM) of order n. Here n equals the number of possible discret grey values of each pixel. The element SDM(i,j) gives the number of pixels with the grey value s(q)=i followed in a distance d(r,ρ) by a pixel with the grey value s(q+d)=j.

$$SDM = \sum_{\underline{q} \varepsilon B} [\, 1 - \sigma \,(\, | s \, (\underline{q} + \underline{d}) - j \, | \,) \, | \, s(\,\underline{q}\,) = i \,] \tag{7-1}$$

The displacement vector d is defined by the absolut difference r und the angle ρ, r and ρ should be selected so that d does not point at interlattice-pixels, that means the multiple of the lattice constant for r and the multiple of 45° f or ρ . It is advisable to limit the used grey values to a number between 4 and 20. If all possible grey levels - 256 in case of a 8-bit-representation - the resulting SDM is of order 256. This leads to a high calculation expence without better results, because small grey value differences are caused mainly by noise terms

The referenced publication describes 14 texture features. Only the first two will be shown here, where p(i,j) denotes the element i,j of the SDM-matrix:

1. Second Moment

$$f_1 = \sum_i \sum_j p(i,j)^2$$

(7-2)

2. Contrast

$$f_2 = \sum_{n=0}^{N_g - 1} n^2 \left[\sum_i \sum_j p(i,j) \right] = \sum_i \sum_j (i - j)^2 \cdot p(i,j)$$
$$\text{with } |i - j| = n$$

(7-3)

Feature f_1 sums the squared elements of the SDM. It has a significant value, if some special grey value transitions appear very often. f_2 sums the Elements of the SDM multiplied by the grey level difference $(i - j)^2$. It is sensitiv for high contrast grey level differences. The other features are sensitiv for special texture effects.

The following example shows the calculation of the SDM (with r=1, $\rho = 0^o$) and the values for the features f_1 and f_2 for two special texture cells s_1 and s_2 :

$$\{s_1(v,\mu)\} = \begin{pmatrix} 1012 \\ 1100 \\ 0210 \\ 0202 \end{pmatrix}$$

$$\{s_2(v,\mu)\} = \begin{pmatrix} 1122 \\ 1122 \\ 1222 \\ 2222 \end{pmatrix}$$

$$SDM_1 = \begin{pmatrix} 113 \\ 311 \\ 110 \end{pmatrix}$$

$$SDM_2 = \begin{pmatrix} 000 \\ 023 \\ 007 \end{pmatrix}$$

$f_1=24 \quad f_2=22 \quad f_1=62 \quad f_2=3$ (7-4)

The distinct reaction of these features to the different texture cells are visible.

With the 4 possible directions and the many possible absolute values of the vector d by calculating the matrix SDM and the number of described features, you get a great number of features. The classification is very expensive in this high dimensional space. On the other hand, so many features are not necessary. The question is, how to select the features which bring the maximum contribution to separate the classes. It is possible to make an empiric selection on the basis of representative test pictures.

An analytic solution for the feature selection using a modified Karhunen-Loeve-Transform (KLT) was published by [TOU et. al. 77]. In the general case with not normally distributed features, the KLT determines uncorrelated but not necessarily independent transformed features. If the KL-transformed features are sorted by decreasing eigenvalues, the first features are the most important to describe the texture. This is the optimal solution to describe the texture in fixed accuracy with a minimum number of features. But here the problem is to distinguish best two classes with a fixed number of features. The above mentioned publication describes a modified transform. The eigenvalues of class 1 $\lambda_i(1)$ and class 2 $\lambda_i(2)$ fit the following relationship:

$$\lambda_i(1) = 1 - \lambda_i(2)$$ (7-5)

that means, the more a feature is important for class 1 the less it is for class 2 and vice versa. So the eigenvalues are a quantity for the efficiency to separate the two classes. For the problem this chapter deals with the modified KL-transform was tested in co-operation with [MANIAS 88].

7.2.2 Classification

We already mentioned above, that it is impossible to estimate expected values for the features of the class background which are typical for the whole background. Therefore it is necessary to design a special classificator. First modified distance classificators were tested. But they did not bring any suitable results. So a special classificator was developed which uses the a-priori-information of the roadway position in the display. The designed classificator interprets rows or columns of texture cells in the scene. For these rows or columns models for the membership of the cells in the distinct classes can be defined. These models are shown in fig. 7-4.

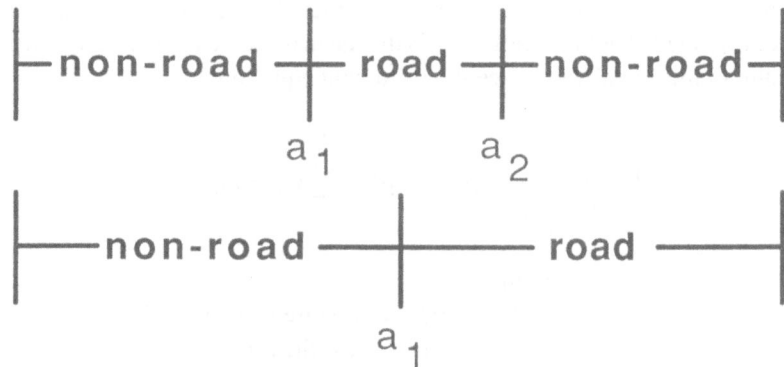

Fig. 7-4 Road model, row- and column-direction

The horizontal (row)-direction starts on the border of the scene with class "background" or "non-road", followed by regions of classes "road" and again "background". To segment this scene, the classificator only has to determine the parameters a_1 and a_2 respectivly a_1 in the vertical case. If one quantizes the distance levels in the model, it is possible to find optimal values for the parameters by minimizing a measure of quality. Fig. 7-5 shows the situation.

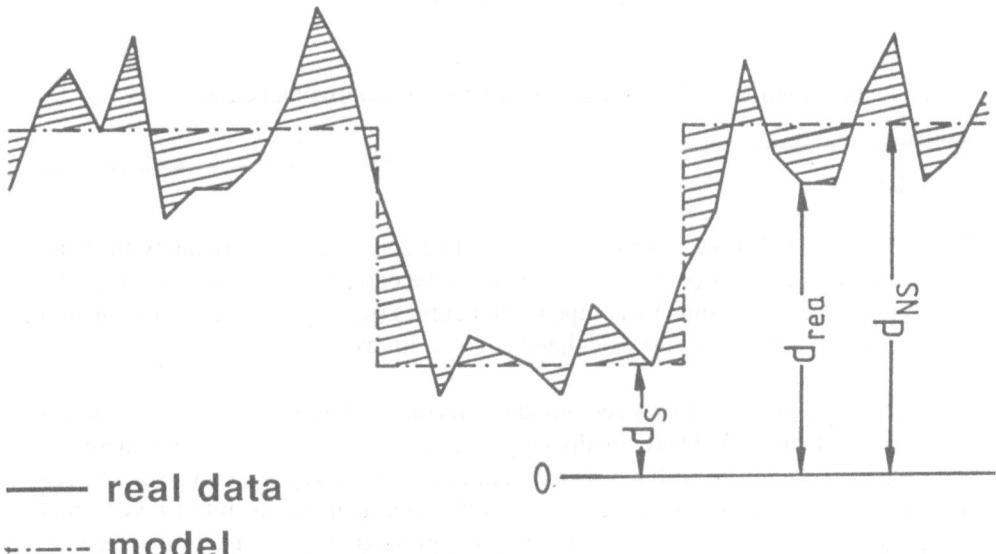

Fig. 7-5   Model - Real Data

The used measure of quality here is the sum of squared differences from the model to the distance d_{rea} of each texture cell. d_{rea} is the mean distance from the expected value of the feature to the actual value (see the following equations).

$$Q = \min_{a_1} \left(\sum_{0}^{a_1} (d_{NS} - d_{rea})^2 + \sum_{a_1}^{m_T} (d_S - d_{rea})^2 \right) \tag{7-6}$$

with
m_T = Number of texture cells per column
d_S = Modelvalue for the difference "road"
d_{NS}= Modelvalue for the difference "non-road"

and

$$d_{rea}^2 = \frac{1}{m_{an}} \sum_{i} \frac{(ds_i - d_i)^2}{ds_i} \tag{7-7}$$

with
d_i = i-th feature in the cell
ds_i = Predicted value for the i-th feature
m_{an} = Number of used features.

An analogous relation with 3 summands exists for the horizontal case.

7.2.3 Results

Fig.7-6 and fig.7-7 show results of the described approach. Unfortunately there are some misclassified cells (see fig.7-6). These can be corrected in a finishing step. One way is to smooth the results by compensation curves (see fig.7-7). The determination of these curves is based on the calculated edge positions.

The described algorithm shows very good results for a couple of test pictures. But, like every approach, it is suitable for distinct types of scenes in a different manner. It utilizes texture brakes, therefore problems exists e. g. in some highway scenes with a concrete road surface enclosed by concrete walls. On the other hand the results are very good for roads enclosed by trees or other regions with significant texture. These are situations where most of the conventional algorithms are unsuitable.

Fig. 7-6   Result 1

Fig. 7-7   Result 2

7.3 Direct edge detection, the PDAF

In the group of direct edge detection algorithms we will consider all approaches which utilize direct edge features like the grey value gradient. Nevertheless, a single global edge detection filter cannot solve the problem sufficiently. It detects edges not only on the road boundaries. There are a great number of detected edges in each scene (see fig.7-8). However, the needed result is a set of coordinates, which describe the road boundary. Therefore global filters can only be a part of the preprocessing. With the available a-priori information in the special case it seems appropriate to use edge following algorithms.

Fig. 7-8 High-pass-filtered scene

Simple edge following algorithms calculate in a first step a prediction point for the next sample period on the basis of previous edge points. The next edge point will be searched only in a neighborhood of this prediction point. For example the "nearest-neigbor-rule" decides for the most likely point in the neigborhood. These algorithms bring missing tracks for most of the complex road scenes.

Better results are obtained by "splitting"-algorithms. If there is more than one track to continue the edge with a minimum possibility, the algorithm branches into a number of tracks. These approaches need additional considerations to cancel senseless tracks and to merge similar tracks. So the calculation time increases.

Another approach calculates the following edge-point as the mean of all possible points in the neigborhood. In contrast to the nearest-neigbor-rule, where only the most probable point is considerd, now all possible points have an effect. In this argumentation the deterministic approaches presented so far meet the PDAF (Probabilistic data association filter), which has its roots in the optimal filter theory. The PDAF considers all points with the a-posteriori-possibility and calculates the optimal following point in the sense of a defined quality criteria.

Which are the common features between the PDAF and the well known Kalman filter? Under some assumptions the Kalman Filter calculates the optimal linear prediction for the state variables with the minimal mean square error. In the subsequent section the parallels of both algorithms will be shown. Nevertheless one cannot find analog optimal properties for the PDAF. The good results will be a verification for this approach.

The Kalman filter algorithm can be decomposed in two parts: The prediction and the correction step. The first step calculates predicted values for the state variables $\underline{x}_{k+1}$ knowing all states until k, the second step corrects these values using the measured output vector at sample period k+1. These facts gave problems in the case of airplane radar tracking. Therefore the PDAF has been developed (see [BAR-SHALOM et. al. 75], [VAN KEUK 85]). If there is more than one airplane in the air space, it is possible to get more than one radar echo. On the other hand, sample steps without any received echo are existing in disturbed situations. If there is more than one echo, one cannot decide which is the right one.

The Kalman filter does not describe these effects. Therefore, the PDAF has two additional parameters: The mean number nf of untrue echos in the gate and the recognition possibility Pd of the correct echo. With nf=0 and Pd=1 the PDAF corresponds to the Kalman filter.

The detection of the road boundaries brings similar problems. Instead of missing or wrong echos you get missing or wrong edge points. With these parallels one can expect that the PDAF will give good results.

7.3.1 The PDAF-algorithm

The following list shows the whole PDAF algorithm. For the derivation see the referred sources.

Model

$$\underline{x}_{k+1} = \underline{F}_k \, \underline{x}_k + \underline{w}_k \tag{7-8}$$

$$\underline{z}_k = \underline{H}_k \, \underline{x}_k + \underline{y}_k \tag{7-9}$$

Parameters of disturbance

$$E\left\{\underline{w}_k\,\underline{w}_j\right\}^t = \underline{Q}_k\,\delta_{kj} \tag{7-10}$$

$$E\left\{\underline{y}_k\,\underline{y}_j^t\right\} = \underline{R}_k\,\delta_{kj} \tag{7-11}$$

One-Step-Prediction $k \rightarrow k+1$

$$\underline{P}_{k+1|k} = \underline{F}_{k+1}\,\underline{P}_{k|k}\,\underline{F}_{k+1}^t + \underline{Q}_{k+1} \tag{7-12}$$

$$\underline{x}_{k+1|k} = \underline{F}_k\,\underline{x}_{k|k} \tag{7-13}$$

$$\underline{z}_{k+1|k} = \underline{H}_{k+1}\,\underline{x}_{k+1|k} \tag{7-14}$$

Correction of the prediction with the last measured value

$$\underline{W}_{k+1} = \underline{P}_{k+1|k}\,\underline{H}_{k+1}^t\,\underline{S}_{k+1}^{-1} \tag{7-15a}$$

$$\underline{S}_{k+1} = \underline{H}_{k+1}\,\underline{P}_{k+1|k}\,\underline{H}_{k+1}^t + \underline{R}_{k+1} \tag{7-15b}$$

$$\underline{P}_{k+1|k+1}^0 = (\underline{E} - \underline{W}_{k+1}\,\underline{H}_{k+1})\,\underline{P}_{k+1|k} \tag{7-16a}$$

$$\underline{P}_{k+1|k+1} = \underline{P}_{k+1|k+1}^0 + \underline{P}_{k+1|k} \tag{7-16b}$$

$$\underline{P}_{k+1} = \underline{W}_{k+1}\left(\sum_{i=1}^{m_k} \beta_{k,i}\,\underline{v}_{k,i}\,\underline{v}_{k,i}^t - \underline{v}_k\,\underline{v}_k^t\right)\underline{W}_{k+1}^t \tag{7-16c}$$

$$E\left(\underline{x}_{k+1}\,|\,\underline{Z}_{k+1},\,\underline{Y}_{k+1\,k}\right) = \underline{x}_{k+1|k+1} = \underline{x}_{k+1|k} + \underline{W}_{k+1}\,\underline{v}_{k+1} \tag{7-17a}$$

$$\underline{v}_k = \sum_{i=1}^{m_K} \beta_{k,i}\,\underline{v}_{k,i} \quad with \quad \underline{v}_{k,j} = \underline{z}_{k,j} - \underline{z}_{k|k-1} \tag{7-17b}$$

Equation (7-8) describes the state variable model, equation (7-9) the measurement equation. In practical tests a third order state model was used. Y and w are unmeasurable white stochastic processes, which are described by the parameters in (7-10) and (7-11). Equations (7-12) to (7-14) describe the one-step-prediction from the sample point k to sample point k+1. These correspond to the analogous equations of the Kalman filter. Equations (7-15) to (7-17) correct the predicted values in regard to the measured value at sample point k+1. The covariance matrix $\underline{P}_{k+1 \mid k+1}$ (7-16) consists of two summands: $\underline{P}^0_{k+1 \mid k+1}$ is the Kalman filter term, the additional echos bring the term $\underline{P}_{k+1}$. $\beta_{k,i}$ are the a-posteriori possibilities for the echo $\underline{v}_{k,i}$, which are the innovations, hence the difference between the predicted values $\underline{z}_{k+1 \mid k}$ and the measured values $\underline{z}_{k+1,i}$. Equation (7-17a) calculates the optimal prediction value.

The great similarity between the PDAF and the Kalman filter can be easily seen in the selected representation. Still missing are the equations to calculate the a-posteriori possibilities $\beta_{k,i}$. For the derivation and complete representation see again [BAR-SHA-LOM et. al. 75] and [VAN KEUK 85].

To limit the calculation effort we make the following assumption (see fig.7-9). The past until sample step k will be described by the probability density function $p(z_{k+1,i} \mid Z^k)$. The assumption is that this probability density function is normal in each sample step:

$$p(z_{k+1,i} \mid Z^k) = NV(z_{k+1 \mid k}, \sigma^2_{zz}) \tag{7-18}$$

The variance σ^2_{zz} will be given by the element P_{zz} of the covariance matrix $\underline{P}_{k+1 \mid k}$ and the measurement error variance σ^2_{rr}:

$$\sigma^2_{zz} = P_{zz} + \sigma^2_{rr} \tag{7-19}$$

Fig.7-9 shows an example. Without the above simplification the resulting density function p* is the superimposition of two functions. A possible consequence should be that the gate, the region where the following edge point lies in with a fixed probability, splits into more than one distinct regions. That brings all disadvantages of the splitting approach shown above.

With the assumptions shown in [BAR-SHALOM et. al. 75] for the distribution of the incorrect echos in the gate it follows for the a-posteriori probabilities:

$$\beta_{k,0} = 1/N_f \, b_k \tag{7-20}$$

$$\beta_{k,i} = 1/N_f \, p(z_{k,i} \mid Z^{k-1}) \tag{7-21}$$

with the scaling factor N_f.

$$N_f = b_k + \sum_{i=1}^{m_k} p(z_{k+1,i} \mid Z^k) \tag{7-22}$$

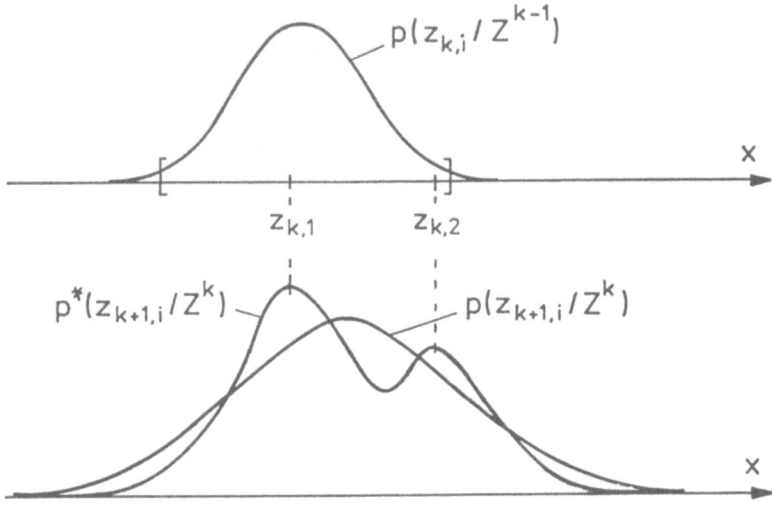

Fig. 7-9 Possibility approach

7.3.2 Adaption to the practical situation

Before using the PDAF a preprocessing step has to detect the possible following points, like the echos in radar tracking. A high pass filter with thresholding brings these discret "echos". Different types of high pass filters were testet without recognizing a significant impact on the following PDAF. The determination of the gradients with optimal calculated compensation planes in small neighborhoods brings better results in the case of high frequency noise parts. Another idea are direction sensitive edge filters. It is a disadvantage that these filters tend to be of higher order. This brings larger calculation times, so that we do not use such filters. As described above, the preprocessing brings a great number of edge points, only a small number of them describe the road boundary. The PDAF will find these.

For the prediction step of the PDAF, a state space model is needed. This model describes the dynamic behaviour of the edge. Here a third order model is used. The state variables are the position x and its first and second derivations v and q, respectively.

$$x_{k+1} = x_k + Tv_k + \frac{1}{2T^2}q_k$$

$$(7\text{-}23)$$

$$v_{k+1}=v_k+Tq_k \qquad\qquad (7\text{-}24)$$

$$q_{k+1} = e^{-\frac{T}{\theta}}\, q_k + \Sigma\sqrt{1-e^{-\frac{2T}{\theta}}}\, u_k \qquad\qquad (7\text{-}25)$$

T is the sampling time, θ defines the correlation between following states (see [VAN KEUK 71]).

7.3.3 Results

The PDAF approach was tested with sequences of real scenes. They were recorded during test drives in real time on a video cassette recorder and then interpreted in the laboratory. The single pictures follow with a frame frequency of 25Hz. The necessary calculation time for each picture is longer than 40ms, therefore the algorithm cannot interpret each picture. This coincides with results in human perception and vehicle dynamic researches that sample periods in the range of 100ms are necessary. The available hardware needs approximatly 200 until 400ms. That was sufficient for the interpreted scenes.

It is a problem to present a sequence of pictures in a book. The presentation of the good results is therefore imperfect. Fig. 7-10 presents a special result by showing 3 single frames. The time lag between the frames is approximatly 5 seconds, the white cross marks the detected edge at any time. The algorithm interprets the broken line without any problems. The same good results are received for a large number of different scenes.

Fig. 7-11 presents the PDAF operation in a different form. Each row corresponds to one sample step. The gate is marked by brackets, the possible following points by "X". To simplify the diagram, the calculated edge points are not marked, they are positioned in the middle of the gate. The diagram shows four different situations. The number of detected possible following points were reduced from situation 1 to 4 by varying a preprocessing parameter. In situation 4 there are possible points only in each fourth or fifth step. Nevertheless the PDAF is able to track the contour. If there is no detected following point in a step, the gate rises.

This uncertainty of the algorithm is also illustrated by fig. 7-12. There the variance of the position estimate is shown for the tracking of an interrupted and a non-interrupted line. The variance increases at the line breaks.

Fig. 7-10 Result for a broken line

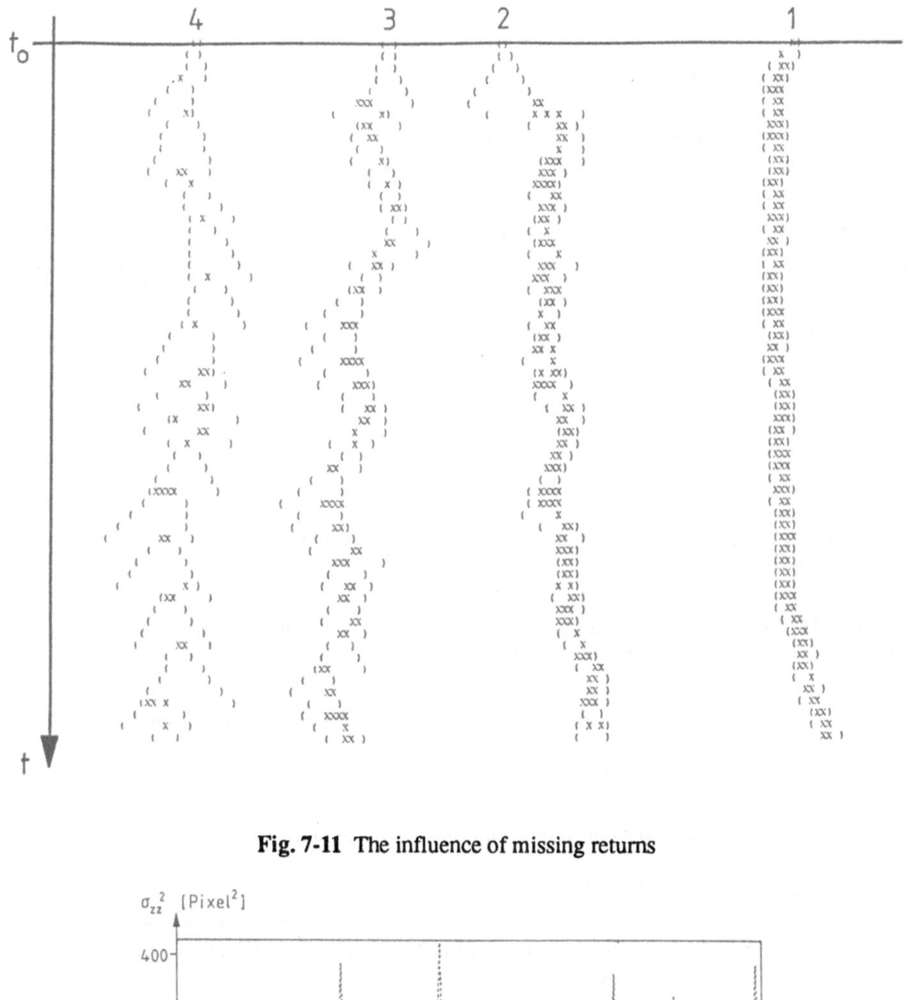

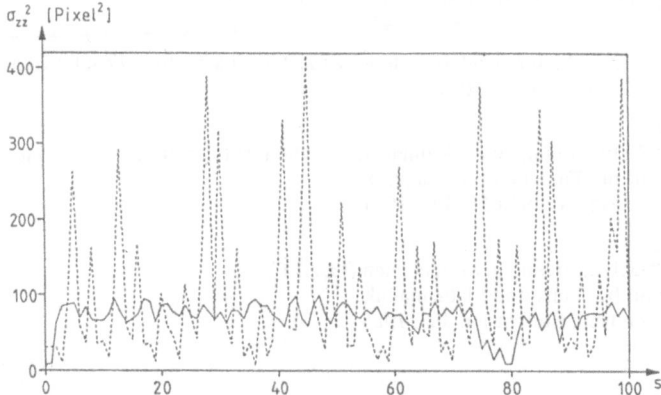

Fig. 7-11 The influence of missing returns

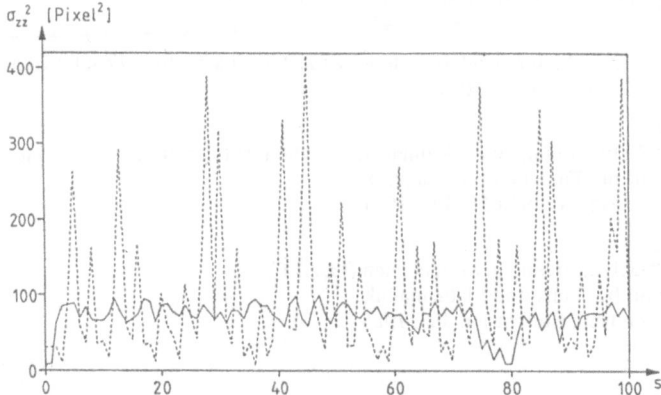

Fig. 7-12 Variance of the position estimate
——— non-interrupted line
- - - - interrupted line

7.3.4 Extension of the approach

The presented approach brings good results for a large number of scenes. But there are problems if the lane edge is masked e. g. by other cars or the edge is not detectable for some other reason. In these scenes the results are improved if the information on the lane width is used. The algorithm has been extended using the estimated lane width. The extended algorithm determines both lane edges. The probability of the following point of one side of the lane has a maximum in the distance of the estimated lane width, measured from the edge of the other side. See [MOEBIUS 88] for a detailed discussion.

7.4 References

[BAR-SHALOM et. al. 75]
Bar-Shalom, Y.; Tse , E: "Tracking in a cluttered environment with probabilistic data association"
Automatica 11, September 1975

[HARALICK et. al. 73]
Haralick, R.M.; Shanmugam, K.; Dinstein, I.: "Textural Features for Image Classification"
IEEE Transactions SMC-3, No. 6, Nov. 1973

[MANIAS 88]
Manias, P.: "Auswahl optimaler Texturmerkmale zur Bestimmung des Straßenverlaufs im Fahrerdisplay" , Studienarbeit
Institut für Regelungstechnik und Systemdynamik, TU Berlin, 1988

[MOEBIUS 88]
Moebius, J.: "Untersuchung zur Spurführung eines Fahrzeugs auf Grundlage der Auswertung des Fahrerdisplays, Dissertation
Institut für Regelungstechnik und Systemdynamik, TU Berlin, 1988

[TOU et. al. 77]
Tou, J.T.; Chang, Y.S.: "Picture Understanding by Machine via Textural Feature Extraction"
IEEE Pattern Recognition - Image Processing, 1977

[VAN KEUK 71]
Van Keuk, G.: "Zielverfolgung nach Kalman-Anwendung auf Elektronisches Radar"
Forschungsinstitut für Funk und Mathemathik,
Wachtberg-Werthhoven, Bericht-Nr. 173, 1971

[VAN KEUK 85]
Van Keuk, G.: "Zielverfolgung in Störgebieten (PDAF)"
Forschungsinstitut für Funk und Mathemathik,
Wachtberg-Werthhoven, Bericht-Nr. 353, 1985

8 Concept of a Multi-Transputer-System and its Application to Parallel Processing

R. Hantsche

8.1 Introduction

The determination of an appropriate underlying computer architecture is an important aspect of building up an autonomous mobil system (AMS). This includes not only the selection of a suitable structure of system-components but also the choice of the components themselves, especially that of the processing element(s). The necessary computing power for such a vehicle - navigating vision-based in a changing environment - lies beyond the performance capabilities of single microprocessors, even if future technology may improve the processing rate further. This is caused mainly by the real-time constraint of an AMS and the complexity of the tasks which have to be solved. Parallel processing, in which computational work is done to a certain extend simultanously, seems to be more convenient to real-time vision and control. This may lead to the desired increase of data-throughput and the opportunity to decompose the AMS-software into more or less independent sub-tasks. For designing a suitable computer architecture besides performance facilities, both other general requirements of a computer architecture and specific conditions of our AMS application has to be considered. Subsequently, there will be made no difference between general demands and application-specific constraints. After discussing the significant requirements, the different architectural concepts and their fundamental characteristics may have investigated in order to find out one which meets the requirements best. The subject of comparing computational models is lengthy and does not match the intention of the book. Therefore, it is not dealt with in detail. Moreover, attention is payed to the key-aspects of parallel processing. Its solutions may also serve as designing principles of a computing system.

It turns out that a homogenous multiprocessor-architecture using the Transputer as a building block comes close to the desired computing testbed. This testbed enables us to aquire experience and develop further solutions to our AMS-approach. The concept and the features of the Transputer considering it's high ability for parallel processing are described next in order to stress the way the Transputer can be employed in an AMS-environment. Some applications which can be regarded as a first step towards development and implementation of more sophisticated algorithms are presented finally. This may give insight into mapping parallel algorithms efficiently to the multi-transputer-system.

8.2 Requirements of an AMS's appropriate computer architecture

The most significant requirements concerning the computer architecture of our AMS-approach are high computational efficiency, the capability of fault-tolerance, incremental expandability, general-purpose facility, minimal design effort and transportability. All of those demands are described in detail in the following:

- High efficiency
  The by far most employed characteristic of a computing system is its performance. This means the computing speed, respectively the data throughput per second. However, the computational speed does not suffice as a criterion alone, it is always related to the cost of the particular system-configuration. Thus only the performance/cost-ratio is a reasonable feature of distinction and should be denoted as efficiency here. The budget of our AMS-computer is limited and should not exceed the cost of a minicomputer. The performance which will be necessary cannot be estimated in advance. It depends on real-time experiments which will be carried out only during the implementation-phase of the project. Therefore the opportunity to adapt the system-power to the actual requirements is demanded.
  The problem is the measurement of the overall performance of a computer system. Usually the computing speed of a monoprocessor-system is given in MIPS (million instructions per second) or MFLOPS (million floating-point operations per second). These rates are evaluated by some minimum group of instructions. So, instead of indicating the performance of the entire system (including I/O-operations), the rates only reflect a minimal execution time of the CPU at best. Moreover, any given user-software may consist of many different instructions. Thus only a rough approximation can be obtained from the above figure of merit. Another often employed standard to measure and compare performances is rendered by special benchmark figures (e.g. Whetstone, Linpack, Livermore). They consist of tuned programs simulating a more or less realistic view of commonly encountered computation in a specific field of applications by including a mixture of essential instructions.
  The performance, however, is a function of many interrelated factors of influence (algorithm, compiler etc.) which cannot be combined into a single performance-value. Therefore, the comparison of architectures by its processing power, regardless of the applied method of measurement, always lacks of general evidence of the architecature's efficiency. The situation becomes even worse comparing parallel architectures instead of sequential working ones, because performance of parallel systems depends on further parameters (the extraction of concurrency, the structure of the hardware etc.). Considering all factors of influence, no general standard measuring the performance of a parallel architecture will be known, but several theoretically derived definitions are collected and discussed in [GONAUSER, MRVA 89]. Despite its insufficiency the MIPS/MFLOPS-rating of a single processor multiplied with the number of processors N (supposing a multiprocessor-sy-

stem) is often found as a performance specification of parallel systems. Thereby an ideal speedup of N is presumed while disregarding all facts of communication and organisation overhead which may cause a performance saturation effect of parallel architectures. A realistic estimation of the expectable speedup employing N processors is limited between the Minsky's assumption of $\log_2 N$ as a lower bound and $N/\ln N$ as an upper bound. The later one regards that the probability of i processors out of N being busy decreases with i [GILOI 81], [GONAUSER, MRVA 89].

- Fault-tolerance
Especially in the domain of man-maschine interference, where maschines are able to jeopardize human's health or even life, the reliability of the computing system has to be asked for. This is true as well in areas, where loss of data or damage of equipment may cause immense costs. Concerning the AMS as well the equipment of the AMS as the environment it operates in should be protected against injury. This damage or loss may be caused by faults ocurring in hard- and/or software. To a certain degree, errors have always to be taken into account. Even in tested software the absence of errors can never be proved and the failure of hardware-components by itself generally cannot be avoided. The only way of dealing with errors is to have effect on the reaction to them.
Reliability in this context means at best that, if any error occurs, the correct execution of a task continues at the expense of a decreasing performance (gracefull degradation). Reliability means at least that a secure state of operation (fail safe) is atained.
To achieve the reliability by means of fault-tolerance generally the following tasks have to be executed [RENNELS 80]:

- fault detection (determination wether a fault in hard- or software has occured),

- fault location (finding the system's component and/or the part of the programm, where the fault has been detected),

- fault containment (avoiding the damage of other parts of the system by error propagation which becomes important in multiprocessor-systems),

- fault diagnosis (identifying the kind of fault) and fault recovery (correcting the fault).

Moreover, the extend to which fault-tolerance is aspired has to be balanced with the delay-time which is introduced by additional routines. If the additional run-time exceeds a certain degree, the AMS may loose its expected real-time-facilities.
To achieve fault-tolerance a redundance in hard- and software is necessary. Because inherent hardware redundancy is provided the distributed computer architectures are superior to centralized systems in the sense of fault-tolerance.

- Expandability
To expand the computing system's facilities (replacing or adding peripheral devices, adding memory, increasing the systems's performance) easy expandability is desired. The most important point is to make the system running faster by only

adding more processing elements without requiring a new design and configuration of the system. A linear increase in processing elements should result in a nearly linear increase in processing power. The necessary system's changes should be as small as possible, because re-designing and re-installing hard- and software has proven to be time-consuming.

- General-purpose facility
  While running a mobile system autonomously a lot of tasks have to be solved methodically. This involves detecting and understanding the vehicle's environment, obstacle recognition and avoidance, road tracking and vehicle's navigation and control. The algorithms which will be used, are not known in advance; partially they depend on experimental results and will change during the project's developing phase. Therefore the computer architecture has to be flexible enough to support approaches of realizing different algorithms. A part of the tasks have to be executed concurrently No unique architecture is equally suitable for all of the above mentioned tasks. . For example, processing images pixelwise (low level processing) require other architectural features than image understanding by symbolic representation of image contents (high level processing). That is the reason why on the one hand only a system of general-purpose microprocessors is able to process the variety of tasks, on the other hand a loss of computing power has to be payed using algorithms which are written in software instead of embedding them in customized hardware like signal processors.

- Minimal design effort
  Due to our situation of staff and finance, it is impossible to design VLSI-devices or chips by ourselves or to develop a completely new architecture. Our approach is hence limited to devices available on the market and to kind of architectures, which have been already employed successfully.

- Transportability
  Using the processing-system as a stand-alone-computer which will be mounted on the AMS a small physicall size, a robustness of assembly and a low electrically power consumption is demanded. These factors can be summarized in the term transportability.

8.3 Key issues of parallel processing and architectural models

Making a decision about the architecture that fits best to our approach of an AMS, we first discussed the necessary requirements. Secondly we have to characterize the possible architectural models and their accompanying features. Up to now one is confronted with the lack of any suitable classification scheme which would allow the comparison of architectures according to all of their different functional and structural properties. The existing taxonomies are only partially able to help the user deciding for an appropriate architecture [GILOI 81], [HOCKNEY, JESSHOPE 81], [GO-

NAUSER, MRVA 89]:

The classification after Flynn determines four classes with two features (number of data - and instruction streams) but cannot make any statement about the arrangement (topology) and the typ of processing elements or about the principle of operation. One of the classes (MISD) is even empty and most parallel architectures can be found in the MIMD-class. Nevertheless, this is a historical scheme mostly used to distinguish between hardware structures. Another taxonomy as explained in [GILOI 81] introduces a lot of abstract architectural concepts by determining a principle of operation (representing the structure of control and data) and a hardware structure (similar to the one of Flynn). This allows a finer degree of classification, but generally does not render more feasible system architectures which already have outgrown an experimental stage. In our case only a few architectural concepts remain for further investigation [HANTSCHE 88/1].

Before giving a brief review of the results which are obtained from comparing the principal features of these architectural concepts, the following key-aspects of parallel processing and their handling are introduced but are also discussed here only in brief:

- The selection of a level of concurrent execution

- Partitioning the algorithm according to the choosen level of concurrency into different single sub-tasks

- Scheduling the sub-tasks for execution to the hardware devices

- Synchronising the running sub-tasks and providing rules for communication between them

- The organisation and access of memory

- The configuration and power of the processing elements and their interconnection-scheme.

According to the underlying architectural principle, these key-problems are stipulated implicitly by the system (e.g. synchronization of a data-flow machine) or handled explicitly by the user (e.g. synchronization with software support in a multi-processor-system).

8.3.1 The level of concurrency

The inherent parallelism of any algorithm may be exploited on different levels of concurrency. The levels can be sorted from fine to coarse grannularity of parallelism in the following manner:

- The operational level, at which elementary logic and arithmetic operations are executed in parallel using multiple hardware devices.

- The instruction level; each instruction consists either of single assignments or of expressions of compounded elementary operations. Multiple instructions can be performed concurrently, if there exists no true dependencies of data between the instructions.

- The process level, at which processes can run in parallel on different processing elements or apparently parallel on a single processing element sharing processor time. Every process contains a collection of sequentially executable instructions, a set of data and a state of activity.

- The user level, at which independent user programms (jobs) are executed concurrently.

Parallelism at operation or instruction level is extracted from a sequential program efficiently only by compilers. The degree of implicit concurrency on this lower (finer) level, however, is limited. The limit is refered to as well data dependencies as conditional and branch statements. A higher degree is possible, if the problem shows explicit parallelism; this means, that the problem includes a structured set of data which can be processed independently (as it is the case with array and vector processing).
Viewing our broad spectrum of parallel tasks to guide an AMS autonomously, the highest degree of parallelism is found at process level, although several processes, for example the sensor inputs of the AMS, produce data sets which are feasible for explicit parallelism at instruction level.
In general, however, the choosen level depends on the structure of application. At present, no way is known to extract parallelism automatically from the process level with no regards to the application. Therefore the user himself is responsible for partitioning the algorithms in parallel entitys. This partly leads to more sophisticated program design compared to the sequential case.

8.3.2 Partitioning

Partitioning means the decomposition of a given computational problem or algorithm at one of the above mentioned levels of concurrency into parallel executable sub-tasks, which are more or less mutually independent. The dependencies refer to necessary data communication.
Graph theoretical techniques can be used to model the problem of partitioning; sub-tasks are represented by nodes and their predecessor-successor relations or their communication structure are indicated by edges connecting the nodes. This mapping provides for automatic detection of parallelism by transforming the graph into a computer-tractable form.

The exploitation of concurrency in existing algorithms is done either implicitly by the compiler checking the instructions against data dependencies or explicitly by the user re-designing the program. Recognizing data dependencies the compiler looks for mutual input and output variables of instructions. Afterwards the sequential program is transformed into a parallel form. The re-design is carried out by applying a language, which contains constructs for the expression of parallelism (as e.g. ParC or Occam). As mentioned before, the implicit automatical partitioning is limited to the lowest two levels of concurrency. At higher levels the user himself has to organize the explicit splitting into sub-tasks .

Expecting high speedups compared to the sequential execution, the exploitation of as much parallelism as possible seems to be suitable. Extracting all possible parallelism results , however, in an great amount of synchronization and communication overhead because of the great number of small subproblems. On the other hand reducing the overhead by clustering sub-problems means to waste parallelism. As well the possibly extractable concurrency as the overhead depends on the granularity of the application within the choosen level of parallelism. For optimal partitioning concerning the speedup, a balance between both has to be found. This is only attained if time profiles of task-execution and overhead are available. Often this is not the case, therefore the values are estimated according to experience and they cause a more or less good partitioning.

8.3.3 Scheduling

The extracted, simultaneously executable sub-tasks have to be assigned to hardware devices for execution. These devices may include besides I/O-devices, memory etc. especially processing elements. Aiming at a minimized usage of resources, the set of tasks $T = \{t1, t2, ..., tm\}$ has to be distributed to the set of processors $P = \{p1, p2, ..., pn\}$. Most often it is wanted to minimize the total run time of the program. The assignment problems only arise to full extend considering a free-configurable ho-mougenous multiprocessor-system. All other architectures imply only a limited varity of assignments, i.e. in a sytem of special purpose hardware-elements, each dedicated task must run on its belonging processor.

The general scheduling-problem is NP-complete, that is the reason why an optimal scheduling policy is intractable. However, for special structures and restricted classes of problems, an optimal solution can be given [BOKHARI 87]. A lot of other strategies as first-in-first-out, round robin, shortest-job-first etc. are known and often applicated. Generally, these strategies do not show optimal property because of their unadaptability to the application structure.

Scheduling is done on the one hand statically by the user during program design or by the compiler at compile time on the other hand dynamically at run time using a scheduling-algorithm. Thus, in the case of statically scheduling the assignment is not changed during the lifetime of the program. To minimize total execution time - assuming more sub-tasks than processors - the work load of processors have to be balanced. This means that no processing element should become idle. Static scheduling does not allow for efficient load balancing, because the runtime-profil of each process often cannot be predicted very well. On the contrary, using dynamic scheduling methods, no scheduling overhead has to be paid at runtime.

Instead of centralized scheduling by a "manager process" or a dedicated hardware chip, self-scheduling is prefered. Centralized scheduling can become a performance bottleneck of the system, whereas self-scheduling provides for load balancing by having every idle processor taken the next executable sub-task out of a queue (task-attraction). This method is only usefull, if the number of working packets are much greater than the number of processors.

8.3.4 Synchronization and communication

A management of operational resources in a parallel architecture is necessary to avoid conflicts accessing shared devices or services or data simultanously. This is provided by synchronisation-methods which coordinate the interaction of concurrent tasks with their environment. According to the mode of communication between tasks different mechanisms are utilized. Communication is performed by passing messages or by using shared variables.

To synchronize the access of common data or the execution of critical sections by more than one task, the principle of mutual exclusion is feasible. This can be realized, for example, by the semaphore-mechanism, the lock/unlock-mechanism or the monitor-construct. On a semaphore S, which is a boolean variable announcing the state of a resource (busy or free), two procedures changing the value of S, P(S) and V(S) can be performed. Attempting to read or write common data, a task has to test and set, if possible, the semaphore S before it is allowed to access the resource. After a successful operation the task releases S. If the semaphore indicates that the resource is already busy, the task has to wait until S changed its value and the resource gets free. If it is necessary to synchronize the succession of access as well, a counter-semaphore can be employed.

While the interprocess-communication is accomplished by message passing, the sychronization between tasks is achieved by means of a protocol. Message passing means that the tasks send and receive data along a communication channel. The behaviour of the two participating processes differs according to the implemented protocol, e.g. "rendezvous"-, "remote-invocation"-, "no-wait-send"-mechanism [GILOI 81]. The process which wants to communicate, sends a request to the considered partner-process, waiting then for the acknowledgement before transmitting the actual message. This protocol is denoted as "rendezvous"-mechanism. If the sending process waits for the acknowledgement after having transmitted the data, the protocoll is called "remote-invocation". With a "no-wait-send"-protocol, no waiting for an acknowledgement is performed; the sending process goes on until it cannot execute further without the requested data or services.

The first two protocols do not utilize the full possible parallelism because waiting-time may be introduced if the communication partners are not ready at the same time. This fact is related to the concept of co-routines which are not organized in a hierarchically manner, and in so far do not match the asymmetricall producer - consumer relationship, at which the producer can provide several consumers with data or services.

8.3.5 Memory access and organization

Most of the communication in a computer system takes place between processor and memory. This often becomes the system-bottleneck. In parallel architectures the bottleneck is caused mainly by access-contention of multiple processors, but it can also be influenced by the divergence of speed between memory and processor. To avoid this divergence memory bandwidth has to be adjusted by carefully determining the cycle time, the modular structure, the number of transfer channels and the access methods.

Generally, memory in parallel systems can be organized as a global shared memory available for every processor or in a distributed manner, thus giving every processor its own local memory. The global memory can also be implemented by splitting it into moduln of memory which are connected by a network. This structure still represents a common logical address-space and partly supports fault-tolerance.

In the sense of reliability and ease of debugging, distributed memory is easier to handle, this is because its local arrangement. On the other hand, system states or scheduling lists cannot be placed in distributed memory-systems globally, but they have to be copied to each local memory. The access-time of distributed memory-systems depends on the distance between the processor requiring the data and the one holding it. On the contrary, shared memory systems show no time dependency of position but of its access contention which, in general, is unpredictable. Therefore, efficiency (this means data throughput) of a distributed memory system is affected more or less due to the memory communication load by its interconnection-scheme, whereas the efficiency of shared memory systems depends on the solution of the occuring access conflicts.

Hybrid forms of memory allocation with both local and shared memory are most suitable for many applications. They give the possibility both to keep important common data in global memory and to put code and data of only local interest into private memory. Hybrid forms are, however, difficult to organize (the problem of data consistency).

8.3.6 Configuration and power of processing elements

The transfer-time of data between processing elements is mainly influenced by the kind of interconnection network. Not only the processing units but also the multiple hardware resources as sensor and actuators have to be addressed through the network. The optimal topology in the sense of minimising the transfer-time, respectively maximising the connectivity, is a point-to-point connection which connects every resource in the system with every other, it is called a complete graph. This is hardly to realize with regard to the effort of costs and hardware ports; the later increase with $O(N^2)$ if the units increase with N. Thus using graph-theoretical approaches, it is attempted to find other connetion schemes with similar connectivity properties but less amount of costs. Since information is then transfered through more than one node, the transfer-time will rise.

Another aspect of topology is the facility of achieving fault-tolerance. If one path in the network is lost because of failure, the goal node should be reachable through other pathes at the expense of slightly degenerated system-performance.

Comparing point-to-point connections modeled by an undirected graph some properties of such a graph can be determined and can help to characterize different networks [AGRAWAL et al. 86], [GONAUSER, MRVA 89]:

-   degree: The number of edges entering(leaving) a node

-   diameter: The longest of all shortest distances between all possible pairs of nodes

-   density: The number of nodes, which belongs to the graph related to its degree and diameter.

-   average distance ad: The average number of connections, which have to be passed between any two nodes in the network (N = number of nodes, d = distance, N_d^i = number of nodes with distance d from node i):

$$ad = \frac{1}{N} \sum_{i=1}^{N} \frac{\sum_{d=1}^{diameter} d*N_d^{i}}{N-1}$$

(8-1)

Networks are also characterized by the way connections are established (*static* or *dynamic* networks) and the manner data transfer is realized (*packet-switching* or *circuit switching*).

Static networks have fixed connections linking the nodes according to the choosen topology, whereas dynamic networks are build up by switches, the setting of which can be changed during run-time.

For packet switching, no direct connections are provided between sender and receiver; messages are rather routed through pathes. Circuit switching means that direct links between message passing participants are established.

A lot of configurations have been proposed and analysed (e.g. 2D-mesh, torus, ring, tree, pyramid, N-hypercube and petersen graph as static networks, bus and cellbased structures as dynamic networks) [UHR 87], [GONAUSER, MRVA 89].

The decision for a specific topology is strongly related to the implemented algorithms, more precisely to the algorithms underlying software graph. A flexible architecture allows to realize its connections by software-switches (dynamic network) in order to be able to adapt to the software requirements even during program execution. Thus, the necessary configuration according to the varying process-arrangements may change within any program. Other advantages of network-reconfigurability are the simplicity of program development and the extended facilities of fault-tolerance. The program development needs no additional knowledge about the underlying hardware-topology and considering fault-tolerance, damaged components or connections can be apparently replaced by arranging the switches in a different manner.

The number of processing elements (PE) employed in an architecture is often inversly proportional to the power of the PE's, that means that on the one hand systems are build up with few (up to 25) general-purpose, high-performance processors, on the other hand architectures are configured as massively parallel systems with more than 1000 elements which are only able to process specific elementary operations; intermediate forms of architectures are also known. In fact, the selection of the number of PEs depends on the user's application; high data parallelism requires simple operations and is therefore suitable for massively parallel systems, whereas sophisticated and nested tasks require more complex operations and thus are mapped to systems with fewer but general-purpose processors. To take advantage of the parallel hardware elements without saturation of speedup, the number of PEs has to be adjusted to the number of parallel sub-tasks.

8.3.7 Review of a few architectural concepts

The first concept of a computing architecture with a sequential working single micro-processor was given by von Neumann. This concept is still used as the underlying principle of a sequential microprocessor-system. As mentioned before, however, a single microprocessor does not meet our requirements, especially concerning the real-time and reliability facilities. To perceive the global environment, for example, a stereo-camera system which will be only one of several sensors of the AMS is employed. Handling the camera data the processor has to be capable of performing two frames of 512*512 pixels with a resolution of 8 bit each every 40 ms (TV-norm). This results in a bytestream of 13,1 Mbytes/s. Therefore, assuming only 100 operations per pixel (derived from human perception) a performance of 1,3 GIPS is required, which is up to now only provided by supercomputers. Moreover, image processing will be only one task to be executed in parallel driving the AMS.

Nevertheless, the von Neumann operational principle can also be found in multipro-cessor-systems. In this case it does not comprise the aggregate system but is the concept of every single processor.

The principle of operation of von Neumann-machines is related to the variable-mech-anism executing every operation on a memory cell or register. This location is referenced by an address and keeps the instruction or data. At lower machine-level, the operands do not show any structure the content can be derived from. This assignment of a meaning happens at the time of interpretation first. The disadvantage of this concept is the permanent access to the memory incuring a system-bottleneck. Instead of significant data mainly addresses of data, which sometimes even refer to another address, are transfered. In addition, the von Neumann concept neither shows any inherent fault-tolerance nor any facility of expansion.

To replace the sequential execution of a single data flow of instructions by multiple dataflows on which operation can take place in parallel, the dataflow is splitted into temporal (*pipelines*) or spatial (*arrays*) dataflows of identical operations (for the exact definition see [GILOI 81]). Pipes and arrays consist of several simple processing elements which are often tuned to the task they are applied to and are ordered as a chain respectively as a 2D-mesh. Synchronisation and instruction control is done by a central command processor.

Pipelines are used for time-sequential data-streams which have to pass several processing stages. The execution of instructions is therefore divided into several phases (e.g. the addition of floating-point numbers is achieved by comparing and adjusting the exponents, by adding the mantisse and by normalising the result). Within each stage the same operation is performed on every element of the data-stream and every element passes the stages one after another. After the pipe has been filled, each cycle produces a new result. The execution time of the stages has to be balanced, because the slowest stage reduces the data throughput. Using N data components, S stages and equal time T for every stage-operation, the achievable gain of concurrency is

$$G = \frac{T_{seriell}}{T_{pipe}} = \frac{N*S*T}{S*T + (N-1)*T} = \frac{N}{1 + (\frac{N-1}{S})} .$$

(8-2)

The central control, the unidirectional coupling and the suitability for special purposes only seems to prevent this architecture from general utilization. Only such applications, at which the task can be structured as regular sequences of operations on data streams are fit for pipelining. Nevertheless, it is a usefull technique for internal architectural components.

Arrays are used for space-ordered data-streams, if the access to nearest-neighbour data is important (e.g. signal processing). The local intelligence of the processing elements which contain only basic arithmetical and logical operations, is very limited as well. Controlled by a single command processor, all PEs execute the same instructions in a lock-step mode. Given P processors, N data packets and S operations each consuming time T, the gain of concurrency is simply best calculated by

$$G = \frac{T_{seriell}}{T_{array}} = \frac{N*S*T}{(\frac{N}{P})*S*T} = P$$

(8-3)

Problem size and array size have to match, otherwise a loss of parallelism and efficiency would occur. This is due to the fact that mapping the data field has to be mapped several times to the hardware array or that only a subset of the PEs is used+. This architectural concept also is restricted to special applications with appropriate ordered data sets which need massively parallel but not very flexible processing (e.g. image preprocessing).

Both, pipes and arrays support parallelism only at the level of operation. Programming will be often neither comfortable nor transparent, because of the accompanying low-level language or because of having to know the underlying hardware. The structural behaviour of both architectures can be simulated, if necessary, on a multiprocessor-system as well. Due to the costs, in this case, however, the multiprocessor-system will contain less number of PEs.

Other considerations as occurring in *dataflow* - and *datatype architectures* are going in the direction of overcoming the deficiencies of von Neumann-machines.

Dataflow-computers avoid the sequential controlflow of von Neumann-machines by forcing the data to control itself. The controlflow is not explicitly stated in the program by instructions. This makes it possible to exploit a fine-grain, implicit parallelism. Each instruction is activated independently by incoming data values, so that its execution starts as soon as all its required input data have arrived. The number of calculating hardware-elements and the true mutual data dependencies are the only bounding factors of parallel operations within dataflow-architectures. Instead of the sequence of instructions, the logical structure of an algorithm, expressable as a data dependency graph, defines the order of execution. If more sophisticated statements as jumps or loops are needed, explicit control-commands have to be added. The commercial feasibility of dataflow-architectures is still under consideration and several difficulties have to be removed. This involves, for example, a suitable programming language and a very high communication bandwidth. Moreover, no other concurrency is possible than that one at operation-level.

Datatype architectures are developed to implement structured data at the hardware-level. Therefore, techniques are used labelling the data with added bits (tagged architectures) or introducing data descriptors (DRAMA = descriptor referenced autonomous memory allocation), which leads to a reduction of administrative overhead, since no effort as in von Neumann-architectures is made to map the structured data to the unstructured hardware and vice versa [GILOI 81].

None of these concepts seems to ensure the development of a general-purpose and parallel architecture which already left the stage of research application and is supported with comfortable software-management and -design tools.

The system architecture which offers most flexibility, is a multiprocessor configuration consisting of at least two processing elements which are as well applicable in a single microprocessor environment and which work autonomously. The principal of operation of each microprocessor is therefore fixed to a sequential von Neumann one. Because of the wide variety of different multiprocessor-architectures, it is advisable to figure out functional and structural possible characteristics of them. At first, the key issues of parallel processing applies especially to multiprocessor-systems (MPS) and can refered to as distinctive features.

Furthermore, multiprocessor-systems are distinguished by the typ of participated processors (*homogenous - inhomogenous*) and the distribution of tasks among them (*symmetrical - asymmetrical*). In a *homogenous* MPS only a unique processor family is used as a building block , whilst in a *inhomogenous* MPS different kinds of processing elements (including special-purpose hardware) are combined. If the distribution of tasks is independent of the processing element, that means if every element is treated equal and is capable to perform any of the tasks, the system is called a *symmetrical* MPS. On the other hand a system with predetermined processors handling special tasks is terme an *asymmetrical* MPS. It becomes obvious, that a symmetrical MPS has to be also homogenous, but the opposite is not true.

Moreover, the kind of interprocessor cooperation (*centralized - distributed*) as well as the degree of interprocessor coupling (*loosely - moderately - tightly*), which implies the way of communication between the units, can help to classify MPSs. In a *centralized* system, a supervisor controls and organizes the processing centrally (master-slave principal) knowing at any time the entire state of the system. The supervisor is considered to be rather a logical software-process which may "float" through the system than a physical processor. If the cooperation is *distributed*, no process or processor exists knowing the overall system's state and therefore no instance is able to assume the control of the system globally. The system-nodes work much more autonomously, thus every processor have to be equiped with a set of basic operating-system's functions and therefore, the only way of communication is message passing. Autonomous complete computer systems belong to the *loosely coupled* systems connected with other systems by serial communication links. On the other hand, *tightly coupled* systems have a shared memory even for programm code and a common operating system. Nevertheless the processors perform their computation individually. Between both the *moderately coupled* MPS consist of processors with local memory, but some periperical resources are shared. The processors work autonomously and the communication takes place on data level.
The above mentioned features show partly redundant attitudes concerning their ability to divide the MPSs into non-overlapping classes: For example a MPS with distributed control would never become a tightly coupled system and a loosely coupled system cannot centralize its control-mechanism.

The maximal expectable gain of an MPS equals the number of processors. Because of transfer, synchronisation and perhaps scheduling overhead, however, a saturation effect which brakes the linear increase of throughput, is introduced while the number of processors is increased. This effect is application dependent and can be weakened by an appropriate partitioning and scheduling and by a proper interconnection-mechanism. Thus the communication load and the access contention is reduced. As mentioned before (see section 8.2) a realistic speedup of N processors lies between $\log_2 N$ and $N/\ln N$ from a theoretical point of view .

Regarding our requirements of an appropriate architecture for the AMS, a homogenous, symmetrical and moderately coupled MPS seems to be the most suitable architecture. It offers together with the Transputer and its facilities which are introduced in the next section an easy expandability in order to increase performance, a hard- and software supported capability of fault-tolerance and an application-independent utilization of the architecture. In the case that dedicated algorithms for the AMS are known, the determined concept may be turned into an inhomogenous, asymmetrical MPS later on, in order to speed up the system using specialized hardware. However, the reliability of the system may decrease, as long as no additional redundance is introduced.

Further aspects dealing with problems and features of an MPS are discussed in [FATHI, KRIEGER 83], [GAJSKI, PEIR 85], [KARP 87], [PATTON 85].

8.4 The Transputer as a building block for a multiprocessor-system

Designing a homogenous multiprocessor-system according to our requirements, a general-purpose microprocessor with suitable features for parallel processing is needed as a building block. The Transputer is a busless microprocessor which was proved to be one of the fastest single processors which has been available at the time of determining this system-architecture. It offers point-to-point connections, hardware primitives for scheduling and easy expandability.

Most of all the 32bit-processor T800 including a on-board floating-point unit, seems to be a powerful element for parallel architectures, although its inherent mode of operation is still based on the von Neumann-principle. The Transputer has up to now only four high-speed serial communication interfaces which are not enough to build arbitrary topologies. They are denoted as links and achieve a maximal transfer-rate of 20 Mbit/s each. By using DMA-ability the data transfer of each link is carried out nearly independently of the CPU and independently of each other link. Synchronisation of the full-duplex capable links is done by a software protocol with three bits overhead, so that an effective netto data rate of 1,8 Mbyte/s unidirectional (2,35 Mbyte/s bidirectional) is achieved. Therefore, the usual problem of bus connection which becomes the system bottleneck because of access conflicts does not appear. Ideally, a linear increase of throughput adding more Transputers should be attainable, if the communication load keeps small compared to the calculation time.

Because the 4 Kbyte memory is on-chip and is accessible in one processor-cycle (50ns given the 20 MHz-version), it shows register-like qualities. This results in a data throughput of up to 80 Mbyte/s for internal memory. Using external memory, supported by a comfortable memory interface, the processor provides a data rate of 26,6 Mbyte/s (four bytes every 150ns/20 MHz). The sustained performance is declared to run up to

10 MIPS and 1,5 MFLOPS (20 MHz). The performance-distance to other 32bit-micro-processor configurations becomes evident, if the results of the Whetstone-benchmark-test are compared (T800 = 4000 KWhetst./s, 80286/80287 = 300 KWhetst./s, 68020/68881 = 755 KWhetst./s) [ECKELMANN 87], [INMOS 88/1]. Despite the fact that modern RISC-architectures (e.g. Intel i860, Motorola 88000) outperform the first Transputer-generation, the Transputer is not obsolete. This is firstly, because the new microprocessors miss the Transputer's ability of easy expansion and consequently its suitability for parallel processing and secondly, because a new Transputer (H1) is announced which can cope with the mentioned chips.

A simple instruction set, a three-step stack register and a few registers for process-control establish the true CPU. The instruction set which incorporates aspects of the RISC philosophy, is build up byte-orientated using four-bit-instructions and four-bit-oper-ands. Though only 16 instructions, the most frequently occuring one's, can be coded in one step. Statistics have shown that 70% of a usual programm contains single byte instructions. The remaining instructions and larger data values are composed of additional four bit pieces which are combined in the operand-register by making use of the "prefix-" and the "operate-instruction". The prefix-instruction loads this register and extends the operand to any length. The operate-instruction is needed to distinguish between operands and instructions by interpreting the contents of the register as an instruction. A compacted code, less memory access of codewords, a uniform representation which is independent of processor-wordlength, and a often one-cycle execution time are the advantages of this principle of instruction generation [INMOS 88/2]. With its two-dimensional block-move instructions which provide a copy-speed of 40 Mbyte/s within internal memory and of 13,3 Mbyte/s using external memory (20 MHz-version), the Transputer obtains powerfull graphic facilities and has not to hide from special graphic processors.
The design of fault-tolerant systems is supported by different signals (ErrorIn, Error, Analyse, Reset) and flags (Error, HaltOnError). They both help to analyse the reason of error by reading out the preserved states of the processor and help to stop faulty processes or processors. Therefore, an error propagation through the system can be avoided. Only one interrupt level which is implemented as an event-channel and takes max. 3,7 µs for response, is provided by the Transputer .

The Transputer gives hardware-implemented support to the software-model of parallel working processes which communicate exclusively by the means of channels, as this is basically considered in the OCCAM-language. So, concerning the real-time aspect, the most efficient way to program the Transputer with a high-level language is the usage of OCCAM, because the direct relationship between this language and the processor leads to a fast and close conversion into machine-code. Every other language has to pay a greater conversion overhead which slows down execution.
With the primitive processes declared in OCCAM, the user gets very flexible to partition his application at several levels of concurrency by using the process as a

software-building block. The basic processes in OCCAM are the input/output-process along channels and the assignment-process. The usable levels reaches from the instruction level (fine granularity), at which every instruction is viewed as a process, up to the task level (coarse granularity), at which every process includes a sequence of arbitrary instructions or processes. So, every process itself may be composed of further nested processes. Programms written in OCCAM according to this process-concept which serve as a tool for structuring and exploiting parallelism, can be fitted directly to the underlying hardware. The hardware configuration may not only be a network of Transputers, but also can consist of just a single Transputer. This is true, because the hardware scheduler allows several processes to be executed apparently parallel on one Transputer by sharing processor time. The direct matching of a soft- and hardware-graph is limited, however, by the number of physical available Transputer-links, which often are exceeded by the number of logical channels between processes. The logical channels, therefore, have to be multiplexed to the hardware channels using a multi-plexer-process between the application and the links. The multiplexer-process runs apparently concurrent to the application-processes.

The microcoded hardware-scheduler implemented in the Transputer shares processor time among parallel processes by applying time slices of maximal 2 ms to each process. This results in a quick process-switching of 1 μs (compared to e.g. 220 μs of the multitasking operation system RTOS-PEARL). On the one hand, the hardware-scheduler indeed removes the need of time-consuming software-scheduling, on the other hand, the scope of scheduling is limited to only one Transputer. Thus, scheduling is not supported for a multi-Transputer-systems; this has to be organized by the user utilizing software-solutions.

Every process is given one of the following states: active (is still being executed), ready (is waiting for execution), passiv (is waiting for input/output or is suspended by the timer), halt (has stopped). To administrate several processes, two linked lists, one for each priority level, exist holding only the processes which are ready and holding the one which is active. Whenever a process is unable to proceed, the next one from the list is taken to become active and the former one is replaced at the end of the list. A process switching occurs if the process having occupied the processor has finished or if one of the following reasons become true: The process has to wait for communication with another process; the process is delayed by the timer; or the process has exceeded its time slice, if it is a low priority process. The deactivation only takes place at certain descheduling instructions, e.g. input/output, jumps etc., in order to avoid the interrupt of the evaluation of expressions. This as well contributes to the fast switching time by only having to save a few processor-states rather than the entire stack of variables of the descheduled process. To install and to organize the process-concept, the Transputer has dedicated instructions for starting and ending processes which take care for adding processes to the scheduling list and not terminating processes before the last parallel component-process has finished. For further details concerning the Transputer-device-architecture refer to [INMOS 88/2].

Care has to be taken by the user, that a high-priority process may not contain endless loops without any input/output or time-delay instruction, because a high priority process is not timesliced as it is done with low-priority ones. Thus, no other process would be enabled to achieve processor time.

Although the implemented scheduling helps the user to design concurrent processes on a single Transputer, the work-load balancing between processes on different Transputers is left to the user. In consequence of the fact, that multiple processes (that means more than one) which are placed on a single Transputer, run only apparently parallel, it seems to be better to distribute further processes to other processors. Despite of this, the user is recommended to put at least two processes on one Transputer in parallel. Otherwise, because of the descheduling at communication instructions, the processor may become idle while waiting for communication.

Communication on the Transputer is achieved by unbuffered message passing along point-to-point connections. It is synchronized with the "rendezvous"-mechanism: A process which wants to input data from or output data to another process, has to announce this partner-process. This is done by covering a special channel variable which indicates the established channel between the two processes. If the receiver is ready, the unbuffered communication is carried out by sending or copying the message, otherwise the sender has to wait until the receiver reaches its corresponding input/output statement. The disadvantage of the "rendezvous"-mechanism (wasting parallelism by waiting) is tempered with the activation of another process at an input/output-requests, so that the processor does not become idle if there is another parallel process at hand. Depending on the location of the two participating processes, either on the same Transputer or on two different ones, the communication is performed along channels which are either implemented as memory words or placed on physical links. With the exception of the channel-assignment, no difference is made, however, in programming the communication of an application. Regarding the communication along physical links, the user supplementary is responsible for assigning the logical channels to physical links by the means of configuration constructs which are added to OCCAM.

The Transputer's high applicability for parallel architectures has been figured out. However, the key-problems of parallel processing have not in all points been solved to satisfaction. This involves limited topologies because of only four links and no dynamic scheduling support for a multi-Transputer-system.
The homogenous, moderately coupled multiprocessor-system with distributed control has been assumed to be the most flexible architectural concept, for which the Transputer is a suitable microprocessor.

8.5 Applications and programming hints

The first step in software-development considering the AMS was to become familiar with the design and the obtainable speedup of parallel programs running on a network of Transputers. Several projects, especially in the field of image processing, were realized. They have allowed to aquire experience and to derive some guidelines how to decompose and implement such algorithms.

The employed hardware-configuration consists of a Parsytec expansion unit Multi-Cluster_1. This is supplied with a frame grabber (TFG) and five boards, each equiped with two Transputers T800. Each T800 has access to 1 MByte of memory. This expansion unit is connected via a serial link to an IBM-AT serving as a terminal and a mass storage device. The development system-environment runs on the host-transputer which is linked to the AT. It as well supports the access to the resources of the AT as provides tools for program-development. The Transputer links are connected statically according to a selected topology by means of plugable flat-cable-connections on the backplane.

A free software-configurable interconnection scheme is aspired next.

8.5.1 Speedup investigation concerning an implemented image processing algorithm

The first project was implemented for testing the achievable performance and efficiency using a quite simple algorithm. By varying the way of implementing the algorithm and by varying several parameters, as the number of processors and the number and size of work packets each network-transputer has to work on, it is hoped to get more insight into the exploitation and the programming of concurrency of a multi-Transputer-system.

In image preprocessing, the problem of having to deal with a large amount of data is faced, which dissipates performance. Often the essential information gets rarely lost if the resolution of the image is reduced in a proper manner. This is one intention of forming pyramidal structures which can be obtained from an algorithm after Burt [BURT 84], called reduce-operator.

The original image is filtered with a mask, the elements of which can be regarded as a discrete approximation of the gaussian function, and simultaneously, the resolution is decreased by a factor of two. The convolution-mask has to fullfill several constraints as to be seperable, normalized, symmetric and gives equal contribution to every pixel:

$$G_k(i,j) = \sum_{m=-2}^{+2} \sum_{n=-2}^{+2} w(m,n) * G_{k-1}(2i+m, 2j+n) \, , \tag{8-4}$$

$$w(m,n) = \hat{w}(m) * \hat{w}(n), \tag{8-4a}$$

$$\hat{w}(m) = \hat{w}(-m), \hat{w}(n) = \hat{w}(-n), \tag{8-4b}$$

$$\sum \hat{w}(m) = \sum \hat{w}(n) = 1, \tag{8-4c}$$

$$\hat{w}(0) = 0.4, \ \hat{w}(1) = 0.25, \ \hat{w}(2) = 0.05. \tag{8-4d}$$

If the process of filtering the output-image of the previous step again is repeated and if the reduced images are put one above the other (taking the original image as the bottom), a pyramid is obtained which has N layers if the image size is $2^N * 2^N$.

Under consideration of the properties of the filter-mask, the computational effort can be reduced. This results in a seperate calculation of rows and columns of the image and some multiplications can be replaced by additions. In order to speed up the algorithm the calculation is performed in integer-arithmetic. Nevertheless, it is implemented in real-arithmetic for the purpose of comparison as well. The weights are converted to $w_{int}(0) = 6$, $w_{int}(1) = 4$ and $w_{int}(2) = 1$ by multiplying everyone with $2^4 = 16$. This allows to use the shift-operation for converting the results back. The most benefit of using integer-arithmetic is gained by being able to take the fast modulo-32 operator TIMES for multiplication which consumes only 4 cycles of processor time instead of 39 as the common multiplication-operator do. To fasten this operation further, the smaller operand has to be placed right hand, because of the internal order of stack organization.

The reduce-operator is implemented in three different ways. This is done in order to study the achievable efficiency and the effect of exploiting parallelism differently. To ease programming in the case of parallel execution, only one level of the entire pyramidal structure is created:

-   Sequential implementation using one Transputer as an image-aquiring device and anotherone as a calculation-processor which gets the image transfered to its local memory and sends it back after having performed the filtering.

-   Parallel implementation configuring a pipeline of Transputers which is used in forward and backward direction. The propagating data-entity consists of a multiple of image lines. This data-entity and the number of working Transputers is interactively adjustable by software.

-   Parallel implementation using a processor-farm which is connected as a tree. The ordered data set is divided into equal sized image-parts, on which an interactively selectable number of Transputers perform the reduce-operator. The size and therefore the number of image-parts are adjustable by software in order to investigate the relation between calculation and communication load. A dynamic scheduling mechanism (task-attraction) handles the attraction of data packets by the network processors and provides therefore work-load balancing.

Sequential implementation

To preserve comparability with parallel realizations, the user-interface and the process of image storing and displaying are separated from the data-transforming process; each is placed on a different Transputer. Thus at first, the image data and a few parameters have to be transfered to the calculation-Transputer. This one performs the reduce-operator and then, sends back the reduced image (to a quarter of size) to the video-memory. The summarized transfer-time amounts to $t_t = 0.23$ µs considering an source-image size of 512*512, which means a transfer rate of 1,4 MByte/s. Measuring the calculation-time results in $t_{cINT} = 1.81$ s and $t_{cREAL} = 2,73$ s. Consequently the integer-version is about a third faster than the calculation with real-arithmetic. The same algorithm was implemented on a microvaxII. There it runs in $t_{cREAL} = 17,9$ s. Thus, with regard to the reduce-operator a speedup of factor 6 is gained from a single Transputer versus a microvaxII. If the size of the image varies, the run-time analysis at different sizes

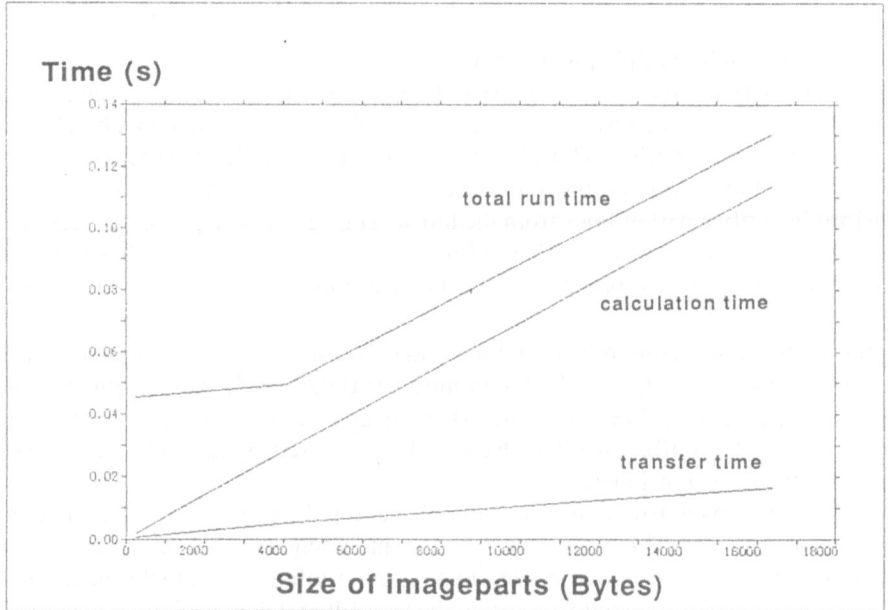

Fig. 8-1 Run-time for the sequential reduce-operator using different image sizes

reveals a linear calculation- and transfer-time above a certain image size (64*64). The overhead nearly disappears for images above this size, whereas for images beneath this size the overhead becomes independent of data size and exceeds the sum of calculation- and transfer-time (see fig. 8-1).

Pipeline implementation

Organizing the MPS as a pipeline (see fig. 8-2) requires a sequential data stream which

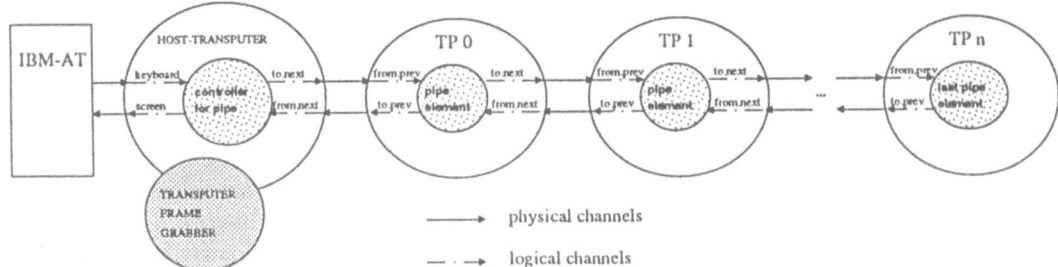

Fig. 8-2 pipeline configuration

has to float through multiple process-stages.

This is originally not the case for filtering the image with the reduce-operator, but the image may be represented by a sequence of multiple lines combined to a block due to the separability of the filter. This allows the filter to be applied first to the rows and then to the columns. Such a block of several lines passes every stage, each of them transforming only a part of lines from the block. This results in equal distributed work load between the stages. The number of lines, on which each processor has to operate, and the number of stages (processors) can be determined interactively by the user.

Supposing the image comprehends b lines and n Transputers are active, each working on m image lines, the pipe is filled with int(b/n*m) data packets of size n*m, where the k-th Transputer calculates the m lines (k-1)*m up to (k*m)-1 from each data packet. The remaining b - int(b/n*m) lines have to be processed by the pipe-master which resides on the host-transputer.

Apart from the image rows, also the columns have to be processed in the same way. That means, after finishing the row filtering including the remaining lines of the host-transputer, the reduce-operator has to be applied column-wise to this intermediate image again. In order to avoid transfering the row-filtered data back to the host-transputer and then using the pipe again, a controller for column-filtering is installed at the end of the pipeline which pumps the data back through the pipe. At the end of the process, the image fully arrives at the host-transputer.

A ring structure, connecting the last pipe-element to the host-transputer, seems to be a better solution, but this configuration prevents from adjusting the number of pipe-elements by software. This is only true as long as no reconfiguration of the network by software-switches is present. A more straight programming and therefore better performance figures will be the result of the software-reconfigurability.

The columens are filtered row-wise in order to allow the use of the same program-code in both directions and in order to access the data according to the linear way they are stored in memory. The intermediate image therefore has to be transposed; this is unfortunately time-consuming.

At configuration time the number of active pipe elements are not known. That is the reason why every pipe-element has to be able to execute both of the two different processes *pipe.element* and *last.pipe.element*. At run-time every pipe-element gets first the number of active elements set by the user. The pipe-element compares it to its own identification number. If both are the same, the main-process branches to *last.pipe.element*, otherwise to the process *pipe.element*.

Within the *pipe-element* two nearly identical processes, which handle the input, calculation and output of data sets, attend to the forward and backward processing of image lines. To utilize the inherent capability of parallelism of a single Transputer, communication and calculation is decoupled. While one data set is filtered, the next one is put in and the former one is put out. Instead of transfering the data sets by channels, data buffers, acting as decoupling devices, are exchanged between the three sub-processes input, calculation and output; thus one stage of an inner pipeline is formed [ATKIN 87]. Since three buffers are necessary, the enclosing process which combines the three sub-processes, is sequentially executed three times. According to the three stages of the pipe, each time another buffer is used; that is, every sub-process within the enclosing process receives a different data buffer as a procedure-parameter which is interchanged at every call.

Varying the number of processors in the pipe shows that a pipeline-structure in the present form is not suitable for implementing the reduce-operator. The underlying partitioning does not really provide a pipeline data flow, but it is made artifically feasible for the pipeline. Because of the missing ability of reconfiguration programing-overhead is introduced additionally. The benefit of decreasing run-time, by involving more processors, is lost already for more than four Transputers. The slope of the speedup even gets slightly negative. This may be caused by the delay times between forward and backward processing and the image-transposition which is always performed on the last pipe-element. Measuring only the run-time in forward direction, the saturation effect is reduced, but does not disappear.

For another measurement the number of image lines m which have to be filtered at every stage are changed, and the number of stages are kept fixed. The image size is always the same, therefore, if m is choosen smaller, the number of data blocks gets higher and vice versa. The measurement indicates that the shortest run-time is achieved by one image line pro stage. This can be also derived from the theoretical gain of a pipeline (see section 8.3.7). If N which is the number of data blocks, runs up to infinity, the gain is maximized. This causes m to become as small as possible, at least m = 1. Furthermore, the measures show that if, on the contrary, the number of lines m are raised, a nearly linear increase of processing time or decrease of gain is given. This increase remains also true for a higher number of stages (processors) in the pipe. Even if lots of small pieces of data are put through the pipeline, as it is the case for m = 1, the expected communication-overhead is obviously not of great weight.

Assigning one transputer to the pipe, the measured run-time with integer-arithmetic is $t_{PIPE} = 1.8$ s. It should be noticed that besides the true pipe-processor a second one has acted as controller and a third one has worked as last pipe-element, so alltogether three processors has been involved. For further results see [HANTSCHE 88/2].

Farm implementation

Another way of processing a group of identical operations on independent data in parallel, is the concept of farming. Farming involves a number of processors executing the same algorithm on different parts of the data. Thereby nearly no interprocess-communication exists except for the distribution of data packets to the network and the collection of results from the network. In the case of the reduce-operator, the image can be divided into sub-images, each being send as a work packet to a net-processor, where the sub-image is filtered seperately and then send back to video memory. A certain communication overhead is incured, because of the local character of the filter requiring also neighbouring pixels. This leads to enlarged data packets with overlapping borders.

For the purpose of runtime-investigation the number of active processors and the size of the sub-images are selectable by software. With the current system configuration, the interconnection-network is not reconfigurable at run-time. A dynamic process scheduler is implemented to balance the work load by the method of task-attraction. The more work packets are send directly to a processor without using intermediate nodes, the higher the possible performance gain will be issued. The work packets are generated within the controller-process. Because of the interface to the user the controller is only able to run on the host-transputer, to which also the video-memory is attached. The controller-process can directly serve at most three more processors; this is true for every processor at further levels as well and is refered to the limited number of links using one link for a connection to the upper layer or to the host.

The entire network is configured as a tree which seems to be suitable for the algorithm, since no interconnection between the processors other than to the controller is necessary. The width of the tree is determined by the number of possible links. In our current system ten Transputers organized within two levels are used for evaluating the reduce-operator. The controller running on the root-node is connected to the three *level1*-transputers and every *level1*-transputer is connected to two *leaf*-transputers (see fig. 8-3).

In addition to the image data, a parameter block containing sub-image size, start-coor-

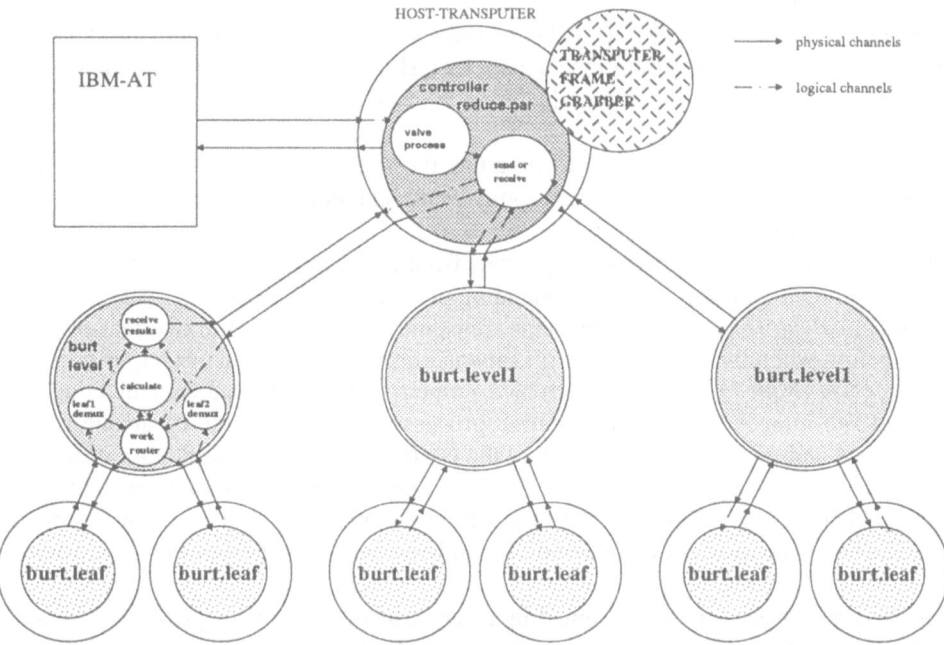

Fig. 8-3 Farm configuration

dinates, elapsed time and sub-image state accompanies every packet to and from the farm. The sub-image state describes different locations of the sub-image within the entire image by determining wether the border of the image is also part of the viewed sub-image. Depending on its state the sub-image is enlarged with additional lines (rows or/and columns) of the original image at the sides, where no border is found. Because the weight matrix (5*5) overlapps at most two pixels in each direction, the highest additional overhead consists of two lines on every side of the sub-image. This increases the communication load for every sub-image of size 128*128 by 6,25% and for every sub-image of size 32*32 by 25%. If the sub-image lies at the border of the original image, the missing image points for the filter-mask at the border-side are created by mirroring the sub-image along this border.

The dynamic scheduler is programmed as a valve process [ATKIN 87] which only allows transfering a new work packet to the processor farm, if the number of idle processors is greater than zero. At the beginning the counter of idle processors is set to the number of transputers which the user has desired to be active. Sending a work packet decrements, and receiving a result packet increments the counter by one. As long as the counter is greater than zero and no result is received, the controller alternatively distributes work packets along the three connected channels to the network. Whenever an incoming result is detected, the channel number is stored and the next sub-image is transfered along this channel to the requesting processor. The system keeps itsself as busy as possible by only distributing work on demand. This can be viewed as a self-scheduling mechanism by task-attraction, although the manner of task-distribution seems to be centralized.

While the level1-processes have to organize the routing of in- and outcoming data packets, the leaf-processes are only responsible for calculation. Within *burt.level1* five processes are working concurrently:
The *work-router* receives the work-packets from the controller, puts it into its buffer and transmits it on request to the demanding processor. The request is made either by the local *calculate*-process on the same transputer or by one of the two leaf-nodes a level below, whereby the local process is given priority if a simultaneous request occurs. The *receive.result*-process is constructed similar to the work-router, collecting the filtered sub-images from the calculation-process of the local node or of one of the leaf-nodes and transfering them back to the controller. The demultiplexing processes *leaf1.demux* and *leaf2.demux* direct the incoming messages over physical links to logical channels according to a tag-byte within the message, either passing results to the receive.result-process or demanding another work packet from the work.router-process.
The *burt.leaf*-process is programmed in a true sequential manner. There is no benefit of further parallelization since the scheduler only sends a new work packet after having received the old one. The parallel execution of communication and calculation becomes only worthwhile, if always two work packets are send on request. This may further improve performance but is not implemented yet. Because shortest communication-time is wanted, all processes transfering data are given higher priority than the ones performing calculations.

When analysing the performance, it will be usefull to think of the elapsed time to be composed of two parts: the time for communication and administration overhead and the true calculation time.
Communication time is measured as the sum of every transfer time to and from the controller process. Transfer time occurring at subsequent levels can not be exploited correctly because it overlapps communication at the higher level.
The calculation time is determined by summarizing the execution time that is needed by the calculation-process on every available processor related to the number of sub-images and multiplied by the number of sequential passes through the network. More than one pass may be necessary, if the number of sub-images exceeds the number of processors. The calculation-time for different passes is assumed to be constant. Thus, one gets an estimation of the distribution of the total run-time into communication and calculation load with N processors.

$$T_{transfer} \approx \sum_{sub-images} T_{send} + T_{receive}$$

(8-5a)

$$T_{calc} \approx \frac{\sum\limits_{N} T_{calc,Tp}}{N} * S$$

(8-5b)

$$S = int(\frac{sub-images}{N}) + \begin{cases} 1, & \text{for } rem\ (sub-images/N) \neq 0 \\ 0, & \text{for } rem\ (sub-images/N) = 0 \end{cases}$$

(8-5c)

The run-time of the reduce-algorithm is measured for all combinations of number of Transputers and size of sub-images. The subdivision of image-size is limited to a power of two for each image direction. The achieved gain, which is defined as $G = T_{seq}/T_{parallel}$, over the number of transputers is showed in fig. 8-4, using the

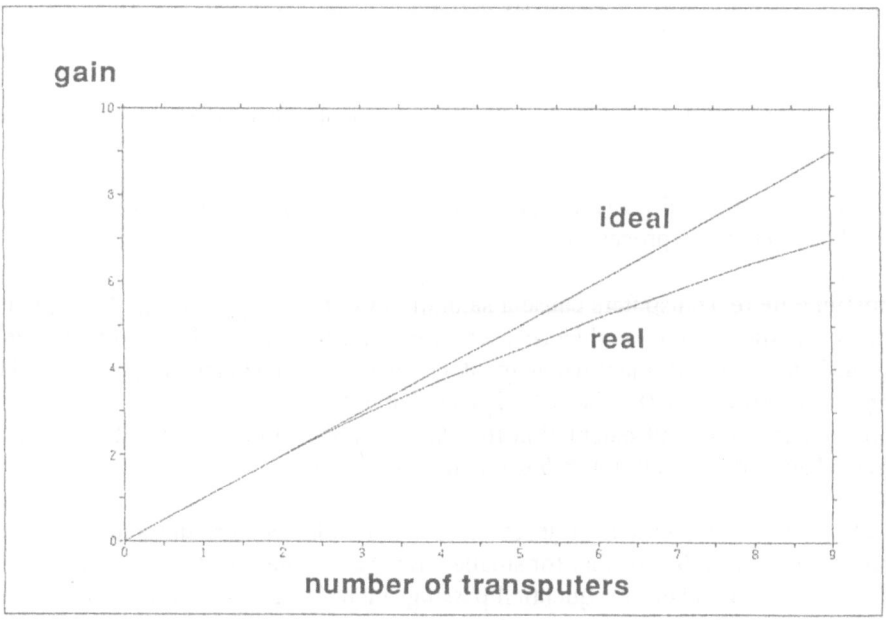

Fig. 8-4 Speedup versus number of transputers using the
farm implementation of the reduce operator

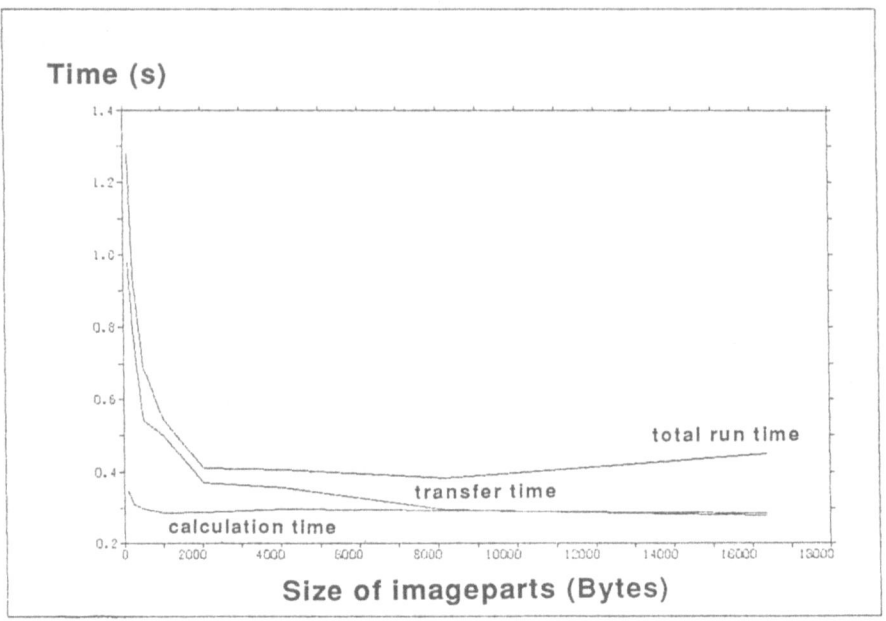

Fig. 8-5 Run-time of 9 Transputers versus the sub-image size

shortest run-time of all possible subdivisions of the image. A linear speedup is only proved for up to three processors.

Employing more Transputers cause a saturation of efficiency. This can be explained with the communication overhead caused by transfering work packets indirectly along intermediate nodes. A maximal gain of 6,9 using 9 transputers is achieved. This expresses an efficiency ($E = T_{seq}/ N*T_{parallel}$) of 77%.
Moreover, transfer- and calculation-time in relation to the size of sub-images are extracted and presented in fig. 8-5 using all 9 processors.

Small sub-images increase the transfer-time remarkable, whereas the calculation-time changes less; this indicates that for smaller size the calculation effort decreases at the same rate as the number of sequentiell passings increases and vice versa. The way the time-rates are evaluated seems to neclect some parallelism, because the total run-time is smaller than the sum of calculation and transfer time. The sub-image size found to be optimal lies between 64*64 and 64*128 resulting in 64 respectively 32 sub-images. The shortest run-time dividing the image into 32 sub-images is $T_{farm,9}$ = 382 ms (integer-arithmetic). For detailed results see [HANTSCHE 88/2].

8.5.2 PDAF implementation

The above mentioned PDAF (see chapter 7) for tracking roadways or indoor pathes according to their boundaries in sequences of real video-images requires real-time processing of the filter. This means that whenever a new frame is put in by the camera, the calculation of the filter should be finished. Thus, within every 40ms (full frame) the edge-point belonging to the path to be followed should be detected. Otherwise it is not possible to analyse every frame. Implementing the PDAF sequentially on one Transputer results in a time-demand of 63,5 ms. This mainly attributes to the frame grabbing itsself. Therefore, to meet the real-time constraint, only the image aquisition and display has to be processed concurrently to the filter. Thus, two processors are necessary, one for handling the filter and the other one for grabbing the image frames and displaying the detected edge points. The filter-calculation includes preprocessing the image window around the last determined edge point in order to find out the possible following edge points, predicting the prospective edge point by means of state model-knowledge and correcting the predicted values in regard to the measured ones

To avoid deadlocks and to organize a secure data communication, a mechanism of non-blocking sending and non-blocking receiving is provided by the introduction of an additional buffer-process. The buffer is realized on an extra Transputer, but also can be implemented on one of the processors which already contain the filter or the frame grabber. This would lead to a buffer-process which then only runs apparently parallel to the other process.

The buffer process guarantees decoupling of the PDAF- and the framegrabber-process. It only stores the actual values of data; image data are transfered from the grabber-process to the PDAF-process, the coordinates of the cursor which indicates the edge-point and the size of the gate, are transfered in the opposite direction. The depth of the buffer is therefore one. This causes at worst a loss of old data, e.g. loss of an image frame, if data are overwritten with new ones before the old ones are fetched by the receiver. This hardly would effect the contour-following because of the slowly changing path-boundaries compared to the image sample time of 40 ms. Moreover, according to the known process-timing values, the clearance of the buffer is always guaranteed before the buffer is filled again. The kernel of the buffer consists of a selective input statement (ALT-construct) which is executed cyclically. Every data transfer can be performed only if an appropriate request of the data-consuming process is sent to the buffer first. An additional condition within the guard of the ALT-construct prevents old data from being transfered twice; this would result in an incorrect edge estimation. The condition is satisfied only, when new data has arrived. Further details and test results are given in [KRUSE 91].

8.5.3 Obstacle detection by using image sequences

An important aspect of guiding a mobile system autonomously, especially on traffic roads, is the avoidance of collision with all kinds of obstacles. To detect these obstacles, a technique based on hypothesis which connects following image-frames, is developed. Two frames from a sequences of images, which are aquired by a CCD-camera, are analysed. Thereby it is pursued to segment the image into areas which are free or even suspected of obstacles. As a first step, a single hypothesis is generated describing the road surface by its geometrical shape. In this way the hypothesis denotes every area, where the hypothesis is not proved to be true, to be suspected of obstacles. To extract suspected areas, the actual image frame and its predecessor are viewed, whereby the predecessor is "distored" by means of the road-surface model giving a prediction of the actual image. The difference between the predicted image and the true, actual image-frame (error location area), reveals areas which do not belong to the road surface and therefore are considered to be suspected of obstacles. To reduce the received, because of simplified assumptions widely spread uncertainty area, the error location mask is combined with a texture analysis of the actual image which detects road and non-road regions. Using contour-based region-growing or region-shrinking methods, only relevant error locations, that means areas that may contain obstacles, remain.
These techniques are implemented in OCCAM on our hardware configuration and discussed in detail in [RITTER 90].
Only a part of the software seems to be feasible for parallel realization. This concerns the prediction of the actual image by distoring its predecessor image pixelwise. Using a plane with an inclined part as model-body of road surface, every pixel of the actual image can be refered to its origin in the predecessor frame by a sequence of transformations of four coordinate-systems. These coordinate-systems are the location-fixed scene, two observer-point-systems and the camera-target-system. Afterwards an interpolation around the estimated pixel-origins of the predecessor give the pixel values of the predicted actual image. This calculation can be applied to every pixel seperately, therefore, the concept of farming is suitable to deal with this kind of parallelism.
An unidirectional, cyclic Transputer-pipeline is utilized. The pipeline is controlled by a master-process running on the host-transputer with access to the video-memory. This proceeding using the pipeline-configuration to farm out the sub-tasks to single pipe-elements, differs from the way the pipeline is applied in section 8.5.1, where really each data packet passes every stage sequentially and thus, only the true pipe-parallelism is exploited.
The master-process is responsible for sending the image data and the parameters to every servant during the initialization phase, and it has to distribute the tasks which have to be solved by the servants to them. Furthermore the master receives the calculated results which have to be stored in the video-memory and generates the stop-signal to finish the pipeline appropriately.
Every servant-process consists of four sub-processes (see fig. 8-6), the *calculator*, the

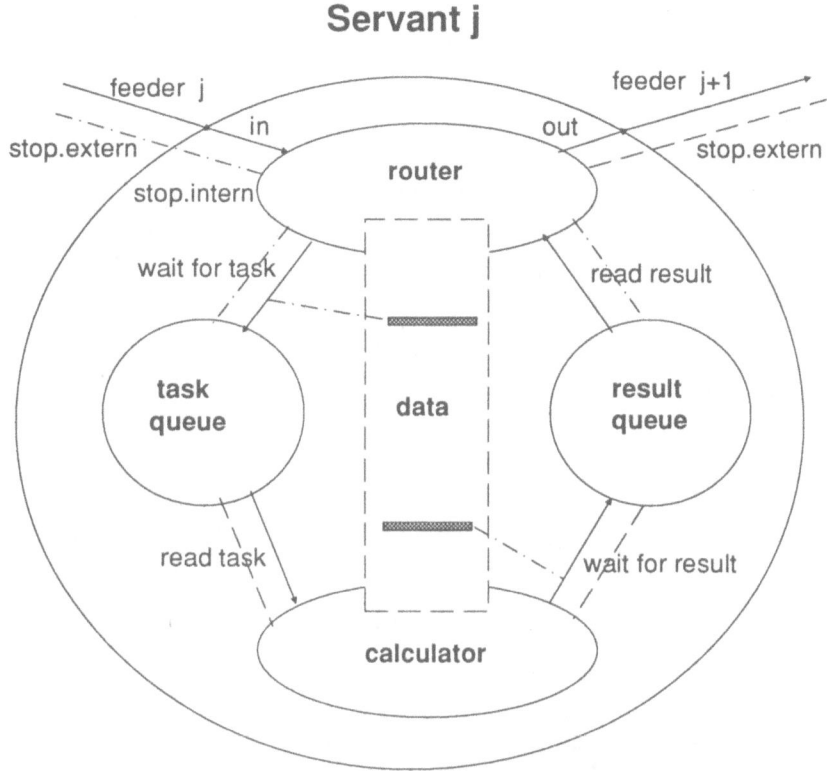

Servant j

feeder j   feeder j+1

stop.extern   in   out   stop.extern

stop.intern   router

wait for task   read result

task
queue   data   result
queue

read task   wait for result

calculator

Fig. 8-6 Process-modell of the servants from [RITTER 90]

router and two *waiting-queues*. All four run in parallel. Each data-direction, that means task-buffering and result-buffering, gets its own waiting-queue. They are inserted between the router and the calculator in order to decouple both processes. Therefore, the router has not to wait until the calculator gets ready for receiving another work packet addressed to it. On the other way round, the calculator can transfer its results back regardless of the state of the router. This would not be the case if a direct communication is implemented between the processes because of the blocked sending- and receiving-mechanism which is used within OCCAM. Each waiting-queue consists of N sequentially connected buffer-processes which receives only one message from the former buffer and sends it to the next one. N denotes the number of possibly delayable messages.

To optimize communication within every servant-Transputer only pointers to data sets are transfered instead of the data itself. The data are stored in a common buffer which is accessible by all processes on the same Transputer. This proceeding just violates the philosophy of OCCAM which says that concurrent processes exclusively communicate

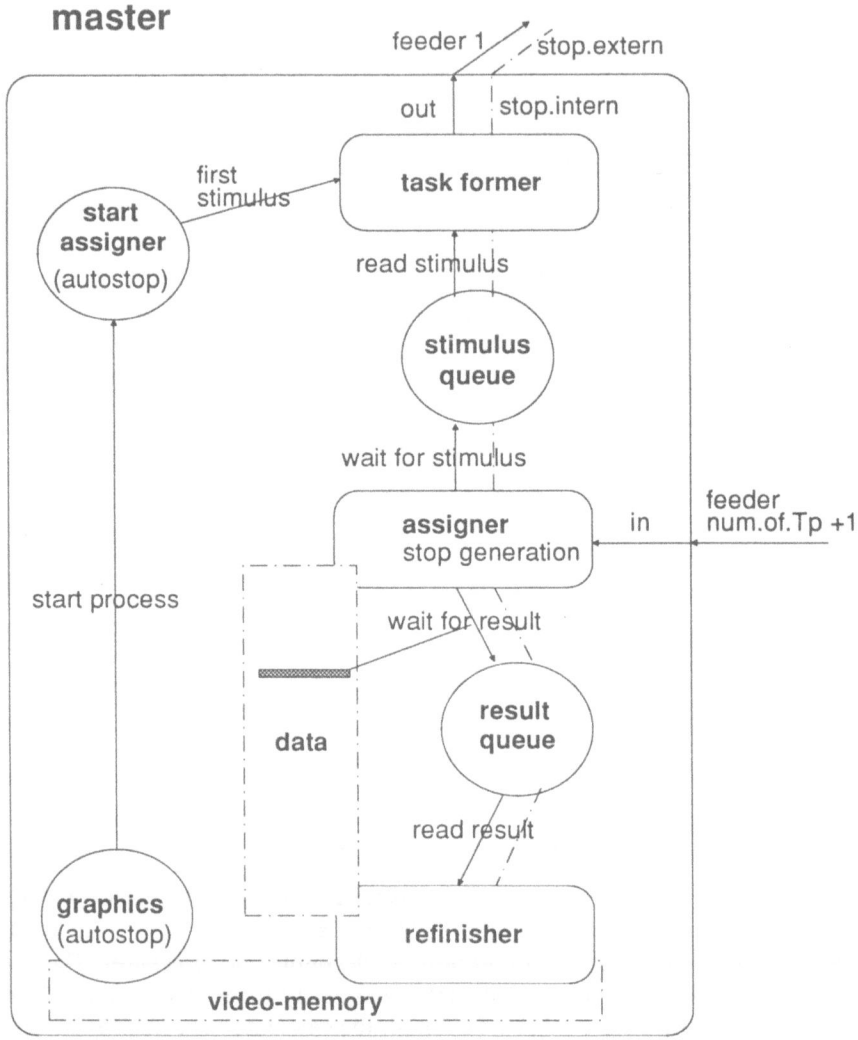

Fig. 8-7 Process modell of the master from [RITTER 90]

by means of message passing along channels. The shared data have to be prevented from being overwritten, before access to the old data may occur. This is avoided by keeping the buffered data unchanged during execution of the servants, since the data are written only once at initialization.

The control of the pipeline is ensured by the master process which combines several parallel processes (refer to fig. 8-7):

The *start-assigner* gives an impulse for the production of a new work packet by generating an address of a goal node and by sending it to the *task-former* which composes two new data sets for the adressed servant and transfers it to the pipe. Every data set consists of a single image line. The adressed pipe-element attracts the work packet, all others let the work packet pass. During the transfer of one result packet back to the master, the calculator-process of the servant is engaged with processing the second work packet.

After having filled the pipe, new work packets are only generated, if a processed data packet has arrived at the *assigner*-process putting it into a global buffer, from where it is prepared by the *refinisher* to be stored at the right location of the video memory. By task-allocation on request and by buffering, a balance of work-load is achieved, keeping every processor busy nearly all the time. Task-allocation on request can be interpreted as scheduling by the method of task-attraction. Since the entire image is completely predicted, the parallel processes are finished properly by routing a stop signal through every process of every servant in order to avoid an undefined system state.

A few applications to investigate a Transputer-based multiprocessor-architecture with regard to our AMS has been presented. Thereby, some problems of parallel processing with Transputers and their handling has been indicated. Generally, it is to say that developing and implementing concurrent programs demands application-dependend solutions. Certain aspects of parallel processing, however, can be found in more than one algorithm and therefore, their handling may serve as a guideline of partitioning, scheduling and implementing similar problems.

8.6 References

[AGRAWAL ET AL. 86]
Agrawal, D. P.; Janakiram, V. K.; Pathak, G. C.: "Evaluating the Performance of Multicomputer Configurations"
Advanced Computer Architectures
IEEE Computer Society Press, 1986

[ATKIN 87]
Atkin, Phil: "Performance Maximisation"
INMOS Technical note 17
INMOS Limited, Bristol, 1987

[BOKHARI 87]
Bokhari, Shahid H.: "Assignment Problems in Parallel and Distributed Computing"
Kluwer Academic Publishers, Boston/Dordrecht/Lancaster, 1987

[BURT 84]
Burt, P. J.: "The Pyramid as a Structure for Efficient Computation"
in: Multiresolution Image Processing and Analysis
Edt.: A. Rosenfeld
Springer-Verlag, 1984, pp 6 - 35

[DIETSCH, ULRICH 87]
Dietsch, H.; Ulrich, R.: "OCCAM - Eine Sprache für die Programmierung paralleler Prozesse"
Informationstechnik it, Nr. 4, 1987, pp. 226 - 234

[ECKELMANN 87]
Eckelmann, Peter: "Transputer der 2. Generation, Teil 1 bis 3"
Elektronik Nr. 18, 19, 20; 1987

[FATHI,KRIEGER 83]
Fathi, Eli T.; Krieger, Moshe: "Multiple Microprocessor Systems: What, Why, When"
IEEE Computer, March 1983, pp. 23 - 32

[GAJSKI ET AL. 85]
Gajski, Daniel D.; Peir, Jih-Kwon: "Essential Issues in Multiprocessor Systems"
IEEE Computer, June 1985, pp. 9 - 27

[GILOI 81]
Giloi, Wolfgang K.: "Rechnerarchitekturen"
Springer-Verlag, 1981

[GONAUSER, MRVA 89]
Gonauser, Monika; Mrva, Michael: "Multiprozessorsysteme"
Springer-Verlag, 1989

[HANTSCHE 88/1]
Hantsche, Rüdiger: "Entwicklung einer Hardwarekonzeption für ein Fahrerassistenzsystem",
Studienarbeit
Institut für Regelungstechnik und Systemdynamik, TU Berlin, 1988

[HANTSCHE 88/2]
Hantsche, Rüdiger: "Erprobung von Algorithmen auf einem Transputersystem im Hinblick auf eine
Parallelverarbeitung", Diplomarbeit
Institut für Regelungstechnik und Systemdynamik, TU Berlin, 1988

[HARP 89]
Harp, Gordon, Edt.: "Transputer Applications"
Computer System Series
Pittmann Publishing, London, 1989

[HOCKNEY,JESSHOPE 81]
Hockney, R.W.; Jesshope, C.R.: "Parallel Computers"
Adam Hilger Ltd., Bristol, 1981

[INMOS 88/1]
INMOS Limited: "Communicating Process Architecture"
Prentice Hall, 1988

[INMOS 88/2]
INMOS Limited: "Transputer Reference Manual"
Prentice Hall, 1988

[KARP 87]
Karp, Alan H.: "Programming for Parallelism"
IEEE Computer, May 1987, pp. 43 - 57

[KRUSE 91]
Kruse, Silko: "Spurführung einer mobilen Plattform durch Auswertung monokularer Bildfolgen",
Studienarbeit
Institut für Regelungstechnik und Systemdynamik, TU Berlin, 1991

[PATTON 85]
Patton, Peter C.: "Multiprocessors: Architecture and Applications"
IEEE Computer, June 1985, pp. 29 - 40

[RENNELS 80]
Rennels, David A.: "Distributed Fault-tolerant Computer Systems"
IEEE Computer, March 1980, pp. 55 - 65

[RITTER 90]
Ritter, Rainer: "Auf Bildfolgen basierendes Hypotheseverfahren zur Hinderniserkennung für
Fahrzeuge", Diplomarbeit
Institut für Regelungstechnik und Systemdynamik, TU Berlin, 1990

[UHR 87]
Uhr, Leonhard: "Multi-Computer Architectures For Artifical Intelligence"
John Wiley & Sons, 1987

Optical Recognition of Chinese Characters

edited by Richard Suchenwirth, Jan Guo,
Irmfried Hartmann, Georg Hincha,
Manfred Krause, and Zheng Zhang

1989. VIII, 144 pp., 73 figs. (Advances in Control System and Signal Processing, Vol. 8; ed. by Irmfried Hartmann) Softcover.
ISBN 978-3-528-06383-2

Contents: Chinese Characters – Properties and Problems: History – Modern Printed Characters – Character Structure - Chinese Characters in the Computer / Input and Preprocessing – Setting the Stage: Optical Input – Picture Segmentation – Size Transformation – Binarization – Edge Smoothing on Binary Patterns / Feature Extraction: Principles of Feature Extraction - Useful Tools – Some Feature Algorithms – Combination of Features – Structural Analysis / Classification: Principles of Classification – Classification Tools – Combining Distance Measures – Hierarchical Classification – The Overlap Problem – Dynamic Classification – Plausibility Checks – Learning Mechanisms / The TECHIS System: Implementation and Results: Hardware and Software Conditions – Program Implementation – Database on Chinese Characters – Tests on Features.

Vieweg Publishing · P. O. Box 58 29 · D-6200 Wiesbaden 1

vieweg